PHILOSOPHICAL DIMENSIONS OF THE NEURO-MEDICAL SCIENCES

PHILOSOPHY AND MEDICINE

Editors:

H. TRISTRAM ENGELHARDT, JR.
University of Texas Medical Branch, Galveston, Tex., U.S.A.

STUART F. SPICKER
University of Connecticut Health Center, Farmington, Conn., U.S.A.

VOLUME 2

PHILOSOPHICAL DIMENSIONS OF THE NEURO-MEDICAL SCIENCES

PROCEEDINGS OF THE SECOND TRANS-DISCIPLINARY SYMPOSIUM
ON PHILOSOPHY AND MEDICINE
HELD AT FARMINGTON, CONNECTICUT, MAY 15-17, 1975

Edited by

STUART F. SPICKER
University of Connecticut Health Center, Farmington, Conn., U.S.A.

and

H. TRISTRAM ENGELHARDT, JR.
University of Texas Medical Branch, Galveston, Tex., U.S.A.

SPRINGER-SCIENCE+BUSINESS MEDIA, B.V.

Library of Congress Cataloging in Publication Data

Trans-disciplinary Symposium on Philosophy and Medicine, 2d, University of Connecticut Health Center, 1975.
Philosophical dimensions of the neuro-medical sciences.

(Philosophy and medicine; v. 2)
Includes bibliographies and index.
1. Neurology-Philosophy–Congresses. 2. Psychiatry-Philosophy–Congresses. 3. Mind and body–Congresses. I. Spicker, Stuart F., 1937– II. Engelhardt, Hugo Tristram, 1941– III. Title. [DNLM: 1. Philosophy, Medical–Congresses. 2. Neurophysiology–Congresses. Wl PH614 v. 2 / W61 T772 1975p]
RC343.T67 1975 616.8′001 76–1204
ISBN 978-94-010-1475-5 ISBN 978-94-010-1473-1 (eBook)
DOI 10.1007/ 978-94-010-1473-1

Originally published by D.Reidel Publishing Company, Dordrecht, Holland in 1976
Softcover reprint of the hardcover 1st edition 1976

TABLE OF CONTENTS

SECTION IV / THE CAUSAL ASPECT OF THE PSYCHO-PHYSICAL PROBLEM: IMPLICATIONS FOR NEURO-MEDICINE

SECTION V / ALTERED AFFECTIVE RESPONSES TO PAIN

SECTION VI / THE FUNCTION OF PHILOSOPHICAL CONCEPTS IN THE NEURO-MEDICAL SCIENCES

INTRODUCTION

> Although the investigation and regulation of the faculties of the human mind appear to be the proper and sole concern of philosophers, you see that they are in some part nevertheless so little foreign to the medical forum that while someone may deny that they are proper to the physician he cannot deny that physicians have the obligation to philosophize.
>
> *Jerome Gaub, De regimine mentis,* IV, 10 ([10], p. 40)

The Second Trans-Disciplinary Symposium on Philosophy and Medicine, whose principal theme was 'Philosophical Dimensions of the Neuro-Medical Sciences,' convened at the University of Connecticut Health Center at the invitation of Robert U. Massey, Dean of the School of Medicine, during May 15, 16, and 17, 1975. The *Proceedings* constitute this volume. At this Symposium we intended to realize sentiments which Sir John Eccles expressed as director of a Study Week of the Pontificia Academia Scientiarum, Città del Vaticano, in the fall of 1964: "Certainly when one comes to a [study]... devoted to brain and mind it is not possible to exclude relations with philosophy" ([5], p. viii). During that study week in 1964, a group of distinguished biomedical and behavioral scientists met under the directorship of Sir John C. Eccles to relate psychology to what Sir John called 'the Neurosciences.' The purpose of that study week was to treat issues concerning the functions of the brain and, in particular, to concentrate upon the relations between brain functions and consciousness. Since the Academy's aim was "to promote the study and progress of the physical, mathematical, and natural sciences and their history" ([5], p. vii), the discussion of philosophical questions was "excluded" by design. Hence, professional philosophers were not invited to participate. Sir John, however, held the view that "all sciences have a philosophical basis and it is generally agreed that there is a philosophy of science which is in fact basic to all scientific investigations and discussions" ([5], pp. vii-viii). Notwithstanding the absence of professional philosophers from the symposium, Sir John maintained at the close of his 'Preface' to the published proceedings of the study week, *Brain and Conscious Experience*, that the study week "was devoted to the greatest scientific and philosophic problem confronting man:

S. F. Spicker and H. T. Engelhardt, Jr. (eds.), Philosophical Dimensions of the Neuro-Medical Sciences, 1–11.

Brain and Conscious Experience" ([5], p. ix). In his 'Introduction' Sir John was compelled to admit that with respect to the Proceedings "the absence of professional philosophers is a ... serious deficiency" ([5], p. xvi). In fact the content of those Proceedings shows that philosophical issues remain alive and unavoidable for neural scientists.

In this Symposium on Philosophy and Medicine we did not take sides with what Eccles called "obscurantist philosophers," those who classify problems relating to brain and mind as pseudo-problems, though it may well be that subsequent to rigorous philosophical analysis the so-called mind-body problem is, in a very specific sense, a pseudo-problem, one which has been traditionally misconstrued in its formulation. The aim of the Second Symposium was to examine some of the central philosophical issues raised by the neuro-medical sciences, and to do so in a context wherein philosophers, physicians, neuroscientists, neurologists, neurosurgeons, and others with interest in the conceptual foundations of the neuromedical sciences could share their vantage points with one another. Unlike the Study Week at the Pontifical Academy of Sciences, the Second Symposium on Philosophy and Medicine was not designed to explore in any detail the mechanisms of neural activity. Instead, the emphasis was to be made in another direction: understanding the significance of neural activity.

The modern neuro-medical sciences developed out of centuries of investigation and debate concerning the function of the nervous system. The problematic of the neuro-medical sciences has not simply involved determining facts concerning neural function but has involved finding a language appropriate to the description of such functions. The availability of both a physical and psychological language for description of neutral functioning insured that the mind-body problem would be a recurring issue in the interpretation of findings concerning the nervous system. One was confronted with the decision of which language is appropriate for what purposes. As it became apparent that the nervous system is not simply another organ system of the body, but is *in some sense* the embodiment of mind, neurophysiology was caught up with the question of how the ways we talk about mind should bear on how we talk about the nervous system. *In some sense*, one experiences and lives in and through one's nervous system, especially the neocortex of the brain, so that such experience can be said to be a function of the nervous system. The problem has been to specify the *sense* in which the nervous system embodies mind or the *sense* in which one lives and experiences in and through one's nervous system, or the *sense* in which mental life can be said to be a function of the nervous system, in addition to specifying the meaning of talk about mental states, such as states of pain.

One classical formulation of the relationship between mental activity and brain function is René Descartes', which provided an account of mind as a unity acting on the body and being acted upon by the body. In his letter of October 6, 1645, Descartes reminds Princess Elisabeth of Bohemia that "In man the brain is also acted on by the soul which has some power to change cerebral impressions just as those impressions in their turn have the power to arouse thoughts which do not depend on the will" ([3], pp. 177-178). Descartes articulated this formulation in his seminal *Treatise on Man:* "... when God will later join a rational soul to this machine, ... He will place its chief seat in the brain and will make its nature such that, according to the different ways in which the entrances of the pores in the internal surface of this brain are opened through the intervention of the nerves, the soul will have different feelings" ([4], pp. 36-37). Earlier, in a letter to Regius (May, 1641), Descartes remarked that the passions, insofar as they affect the mind, have their seat in the brain, "since only the brain can directly act upon the mind" ([13], p. 103).

Specifying the mechanism which accounts for the interaction of mind and brain at the locus of the pineal gland, Descartes says that "...among these figures [i.e., what can cause the soul to sense qualities]... only those traced in spirits on the surface of gland *H* [the pineal], *where the seat of imagination and common sense is,...* should be taken to be ideas, that is to say, to be the forms or images that the rational soul will consider directly when, being united to this machine, it will imagine or will sense any object" ([4], p. 86). A bit further on he adds: "And this will permit the soul, when there will be one in this machine, to sense different objects through the mediation of the sense organs similarly arranged, no change occurring other than in the situation of gland *H* [the pineal]" ([4], p. 95). Descartes thus afforded a view of the human condition that allowed the body to be treated as a machine, and the mind as a distinct and potentially separable immortal soul. This vision was one that forwarded anatomical investigations of the structure of the body machine without directly challenging theological concerns about the operator of the machine. Sometime in August, 1641, in a letter to Hyperaspistes, Descartes asserted with certainty that he had "no doubt that if [the mind] were taken out of the prison of the body it would find [self-evident truths] within itself" ([13], p. 111). The view that the body machine could be understood purely in its own terms became widespread, and was so all-pervasive in the 17th century that it was (and still is) misconstrued as a view shared by the physician, Julien Offray de la Mettrie, the author of *L'Homme Machine* [14].

Descartes' account, though, was freighted with difficulties. As he himself admitted, "It does not appear to me that it is possible that the human intellect can at the same time very distinctly conceive of the distinctions of mind and body and their union. That is the case, for to conceive of them as a single thing and at the same time to conceive of them as two is a contradiction" ([2], p. 691). Descartes simply asserted that mind and body were related in one structure of the brain. Consequently, he did not map mental functions onto the brain; the mind acted as a unity on the brain. After all, for him mind and body were two things, mysteriously, i.e., divinely, related through one part of the body. Thus his attempt, which is often taken as formulating the modern problematic of the mind-body relation, collapsed on its conceptual base in omitting an account of how mind and body are related.

A poignant and classic criticism of Descartes' inability to provide anything more than a bald assertion of a relation was given by Benedict Spinoza, whose criticism bore on the lack of explanatory force in Descartes' account. "[Descartes] affirms that the soul or mind is united specially to a certain part of the brain called the pineal gland, which the mind by the mere exercise of the will is able to move in different ways, and by whose help the mind perceives all the movements which are excited in the body and external objects" ([18], p. 252). "What does he understand, I ask, by the union of the mind and body? What clear and distinct conception has he of thought intimately connected with a certain small portion of matter" ([18], p. 254)? Descartes' inadequate response appears in a letter of August, 1641 to Hyperaspistes: "We know by experience that our minds are so clearly joined to our bodies as to be almost always acted upon by them; and though in an adult and healthy body the mind enjoys some liberty to think of other things than those prescribed by the senses, we know that there is not the same liberty in those who are sick..." ([13], p. 111).

Spinoza's criticism remains to the point with respect to modern attempts to talk of the brain as locating the mind, or attempts to talk of the mind as acting on or being acted on by the brain: how could such disparate entities interact? In what sense is the mind an entity? In particular, if one speaks of the brain's functions in mental terms, what does that talk mean? Does it imply that the mind is identical with the brain? Descartes' answer is that it does not. In another letter to Mersenne, dated April 1, 1640, Descartes observed that besides a kind of recollection in which the mind is said to turn and elicit memory "traces" previously "impressed on the brain," ([3], p. 211), he believes "there is also another kind of memory, entirely intellectual, which depends on the soul alone" ([13], p. 72). In addition to "the folds of memory" – clearly an allusion to the convolutions of the cortex – Descartes reminds Mersenne (August 6, 1640) that there exists "another sort" of memory which is not bodily memory "whose impressions can be explained by these folds in the brain," but is in the intellect, "which is altogether spiritual," and not found in animals ([13], p. 76). Thus brain activity is not, for Descartes, a necessary condition for all forms of human memory.

In his reply to a fifth set of objections to his *Meditations on First Philosophy* by P. Gassendi, whose arguments Descartes refutes as mere "cavillings" and "carpings that require no answer" ([3], p. 211), he remarks that there is nothing at all common to thought and extension. He then adds: "I have often also shown distinctly that mind can act independently of the brain; for certainly the brain can be of no use in pure thought; its only use is for imagining and perceiving" ([3], p. 212). Although neurophysiological processes are admitted, that alone does not justify the claim that they exist when the pure understanding is active. That is, acts of pure thought do not require neurophysiological events. The human intellect is, for Descartes, independent of brain activity; the brain is not involved in thought or pure understanding; reason has no corporeal basis. Simpler arguments, drawing explicitly on Descartes, are to be found in the works of the 19th century physiologist, Marie Jean Pierre Flourens [8]. Descartes' influence was profound.

The radical dualism of Descartes is, however, rejected by many neurophysiologists, though not all. But escape from that dualism is difficult. On the one hand, if one does not have a non-mentalistic language for the description of brain function, one is involved in the mind-body problem as soon as one specifies particular brain functions. For example, to say, as does Wilder Penfield, that "stimulation activates a neuron sequence that constitutes the record of the stream of consciousness" ([15], p. 233) is to require an examination of philosophical issues. What do neurons have to do with the stream of consciousness? And if one does describe neural functions in non-mentalistic terms as, for example, Hughlings Jackson suggested, how is one to talk of the relation that appears to obtain between conscious life and the functioning of the brain? In short, the problematic seems real but unresolvable if formulated in terms of relating mind and body as two separate entities of radically different characters.

One of the first attempts to reformulate the problematic in a non-metaphysical fashion so as to escape these difficulties was offered by Immanuel Kant. Kant gave his position as part of his general theory of knowledge and in his criticism of neuroanatomist, Samuel Thomas von Soemmerring's *Über das Organ der Seele* [17]. Kant argued that talk of locating mental functions (i.e., descriptions of inner sense) in the brain (i.e., in terms of descriptions of experiences through outer sense) cannot in principle succeed if one means by that relating the mind as a thing to the brain as a thing. Kant held that mind and body are not two things, but two ways of experiencing the world ([11], p. 359, A 392). "Thus one can perceive the soul only through inner sense, and the body (either internally or externally) only through outer sense" ([12], p. 35). This general criticism was brought to bear by Kant on the neurophysiological account of Soemmerring in which it was asserted that the *sensorium commune* was located in the fluid of the ventricles of the brain. In responding to Soemmerring, who asked to dedicate his book to Kant, Kant reflected: "Most people believe that they feel their thought in their head, but that is a mistake based on a subreption – namely, they confuse the judgment that the cause of sensation is at a certain place (the brain) with the experience of the cause being at that place" ([12], p. 34). Kant was making the point that even if certain mental functions require certain physiological functions and anatomical structures, it does not follow that the mind (i.e., the soul) is a thing located in those structures. One can correlate functions, but such correlations do not allow any metaphysical conclusions.

But neurophysiologists have undertaken a mapping of mental functions onto brain structures, which mapping has invited conclusions concerning the identity of mind and body, or at least concerning the concomitance of mind and body. Consider the remarks made concerning the relation of mind and brain by Korbinien Brodmann after whom the areas most frequently used in mapping cerebral functions are named: "The difference, the gradations of kind and degree in higher conscious processes are, accordingly, only the expression of an infinitely great variability in the functional summation of individual cortical organs" ([1], p. 303). That is, a great deal of the history of modern neurophysiology involved a movement from an interactionist model to one of an identity theory or a parallelism. To take an example of a parallelist account, David Ferrier, one of the major contributors to the modern neuro-medical sciences, held that: "In accordance with this position it must follow, from the constitution of the cerebral hemispheres, that mental operations in the last analysis must be merely the subjective side of sensory and motor substrata..." ([7], pp. 425-426).

But interaction language persisted in many accounts; consider Theodor Fritsch and Edward Hitzig's remark in their famous article on the electrical excitability of the cerebrum. "To characterize this condition more precisely, one might perhaps put it this way: there was some motor connection from the soul to the muscle, while there was some break in the connection from the muscle to the soul" ([9], p. 331). In short, the neural sciences have appeared to be inescapably bound up with philosophical issues concerning the mind-body relationship and the philosophy of mind generally. Moreover, these issues concerning the nature of the relation of mind and body have continued to be discussed by neurophysiologists [6, 16].

This very partial sketch of some of the debate concerning the philosophical dimensions of issues in the neuro-medical sciences should indicate that the philosophical issues to which this Symposium was addressed grow out of a long history. This Symposium represents a continuation of this ongoing dialogue. It is an attempt to achieve further clarity concerning what one can say about states of minds and about their relationship to their embodiment, as well as an attempt to understand the significance of certain therapies which involve the central nervous system. If the central nervous system is in some sense the embodiment of mind, then understanding that sense will be of particular importance in appreciating the human condition, in understanding what it means to be in the world. However the neural sciences proceed, their directions will both presuppose and shape decisions about what can be said about neural functions. In short, philosophers and neural scientists are joined through a common set of interests in understanding mental functions and their bearing on neural functions.

The first set of papers addresses the recent conceptual history of the neural sciences. William Bynum's paper offers a sketch of the role of Cartesian ideas in disputes in early 19th century neurophysiology that turned on the extent to which mental functions can be mapped onto the brain and the extent to which neurophysiology can be pursued through a non-mentalistic language. The status of mind in these disputes also bore on the extent to which human neurophysiology could be seen as of one fabric with non-human neurophysiology, an important issue for comparative anatomy and physiology and for the significance of generalizations from work on animals to conclusions concerning human anatomy and physiology.

Arthur Benton offers a discussion of the importance of the concept of hemispheric dominance and its role in the development of a language for localizing mental functions. In addition to the dualism of mind and body, the issue of hemispheric dominance offers a compounding problem: the duality of brain functions in the cerebral hemispheres. Further, this is an issue which bears upon the significance of surgical procedures which split the corpus callosum, raising the issue of whether the person involved is thus split into two persons, or whether there is an underlying basic duality in mental functions. These theoretical issues consequently have implications for our understanding of the meaning and morality of surgical procedures directed against the nervous system. For example, if sectioning the corpus callosum produces two minds in one body, special moral concerns about bisectioning the brain may arise.

Further, psychosurgical procedures raise moral issues because they bear on our very presence in the world. That is, psychosurgery seems especially fraught with philosophical and, in particular, ethical concerns because it involves the core of our basis for thinking, feeling, and acting in the world: the core of our embodiment – the brain. Moreover, psychosurgery affords a way of doing physical violence against our minds by altering the structure of our embodiment. In addition, psychosurgery offers an opportunity of enforcing social judgments under the rubric of medical treatment. In this volume, the general question of the moral problematic of psychosurgery is examined by Joseph Margolis and Jerry Fodor in a pair of contrasting analyses.

The papers of Karl Pribram, Marjorie Grene, and Hubert Dreyfus represent an interchange concerning the status of the mind-body relation and the meaning of talk about mental functions in the context of the neural sciences. These papers by a neural scientist and by two philosophers led to an intense debate which was resumed in part of the Round Table Discussion, and is published at the end of this volume. These three quite non-Cartesian discussions are followed with a defense by Hans Jonas of a somewhat Cartesian psycho-physical interactionism by which he gives a presentation of the conditions for free will. Again, basic philosophical issues concerning mind and body bear on ethics and our general understanding of the human condition.

The final set of papers addresses the meaning of pain and in particular altered affective responses to pain. For example, pain appears to be appreciated paradoxically in persons who have had prefrontal lobotomies. The pain is reported to be present though the person experiencing the pain is no longer concerned about the pain. An unbearable pain becomes bearable, not because it is diminished, but because the anxiety, the affect of concern that usually accompanies pain, is no longer present. The pain ceases to be painful in the sense of provoking a need to escape the stimulus. Sinmilarly, phenomena such as masochism involve a seemingly contradictory state in which a pain is desired. The papers by George Pitcher, David Bakan, Bernard Tursky, and Jerome Shaffer analyze the meaning of pain from empirical and conceptual points of departure, offering a wide range of explorations of interest to philosophers, scientists, and clinicians. In particular, the overlap of empirical and conceptual issues gives an illustration of the role of philosophical analysis in a medical context, and the role of the medical context in giving a place for philosophical investigation.

The last section of the book comprises the result of a Round Table Discussion held as the closing session of the Symposium. In short, the discussion ranges from theoretical and historical issues to their role in medical practice. This broad range of issues concerning conceptual questions in the neuro-medical sciences and in clinical practice sustains the point of this collection of essays: conceptual issues are ingredient in neurophysiology, neurology, and neurosurgery. They can be best treated only by acknowledging their central philosophical dimensions. To do them justice, to come to terms with the meaning of mental states and their relations to phenomena studied by the neuro-medical sciences, requires doing philosophy. Moreover, the neuro-medical sciences have much to offer philosophy: conceptual issues with consequences bearing on how research and practice will be understood. The interchange is thus one which can contribute to both philosophy and medicine.

This volume represents the proceedings of the second in a series of symposia concerning philosophy and medicine, the first of which was held at the University of Texas Medical Branch, Galveston, Texas, in May, 1974, with the theme, 'Evaluation and Explanation in the Biomedical Sciences.' The Third Symposium will be held December 11, 12, and 13, 1975, at the University of Connecticut Health Center in Farmington, with the theme, 'Philosophical Medical Ethics.' The Fourth Symposium, to be held at the University of Texas Medical Branch, May 16, 17, and 18, 1976, will resume the examination of a number of themes touched upon in the Second Symposium and in particular will be concerned with the concepts of mental health and disease under the title, 'Mental Health: Philosophical Perspectives.' Future symposia will deal with yet other conceptual issues in medicine. It is hoped that this series of symposia, and the volumes in the series, Philosophy and Medicine, which will result from them, will stimulate sustained trans-disciplinary interest for the many mutual concerns in philosophy and medicine.

The editors take this occasion to express their gratitude to the Rockefeller Brothers Fund, and the School of Medicine of the University of Connecticut Health Center for their support of the Second Trans-Disciplinary Symposium on Philosophy and Medicine held at Farmington. We especially wish to acknowledge our indebtedness to Carolyn C. Brinzey, Stanley Ingman, Ian R. Lawson, Robert U. Massey, John W. Patterson, Edmund D. Pellegrino, Melville P. Roberts, and James E.C. Walker, among many others, who contributed generously in numerous and important ways to the realization of this Symposium. In particular, the editors wish to register their gratitude to Mary Beth Krafcik, whose meticulous devotion to and indefatigable labors with the Second Symposium were indispensable to its existence.

STUART F. SPICKER
H. TRISTRAM ENGELHARDT, JR.

July, 1975

BIBLIOGRAPHY

1. Brodmann, K.: 1909, *Vergleichende Lokalisationslehre der Grosshirnrinde*, J.A. Barth, Leipzig.
2. Descartes, R.: 1899, *Oeuvres*, Vol. 3, *Correspondence*, Descartes à Elisabeth, June 28, 1643, L. Cerf, Paris.
3. Descartes, R.: 1967, *The Philosophical Works of Descartes*, Vol. II (transl. by E.S. Haldane and G.R.T. Ross), Cambridge University Press, Cambridge.
4. Descartes, R.: 1972, *Treatise on Man*, (transl. by T.S. Hall), Harvard University Press, Cambridge, Mass.
5. Eccles, J.C. (ed): 1966, *Brain and Conscious Experience*, Springer-Verlag, New York.
6. Eccles, J.C.: 1970, *Facing Reality*, Springer-Verlag, New York, especially pp. 118-134.
7. Ferrier, D.: 1886, *The Function of the Brain*, G.P. Putnam's Sons, New York.
8. Flourens, M.J.P.: 1845, *Examen de la Phrénologie*, 2nd ed., Paulin, Paris.
9. Fritsch, G. and E. Hitzig: 1870, 'Ueber die elektrische Erregbarkeit des Grosshirns', *Archiv für Anatomie, Physiologie und wissenschaftlichen Medicin*, pp. 330-332.
10. Gaubius, H.D.: 1747, *Sermo academicus de regimine mentis quod medicorum est,* Balduinum van der Aa, Leyden; transl. in Rather, L.J.: 1965, *Mind and Body in Eighteenth Century Medicine*, University of California Press, Berkeley and Los Angeles.
11. Kant, I.: 1968, *Critique of Pure Reason* (transl. by N.K. Smith), Macmillan, London.
12. Kant, I.: 1922, 'An Samual Thomas Soemmerring, August 10, 1795', in *Kants gesammelte Schriften*, Vol. 12, *Briefwechsel, 1795-1803,* Walter de Gruyter, Berlin.
13. Kenny, A. (ed.): 1970, *Descartes Philosophical Letters,* Clarendon Press, Oxford.
14. La Mettrie, J.O. de: 1748, *L'Homme Machine*, Zucac, Leyden.
15. Penfield, W.: 1966, 'Speech, Perception and the Uncommitted Cortex', in J. Eccles (ed.), *Brain and Conscious Experience*.
16. Sherrington, C.: 1963, *Man on His Nature*, Cambridge University, Cambridge, especially pp. 208-220.
17. Soemmerring, S.T.: 1796, *Über das Organ der Seele,* F. Nicolovius, Königsberg.
18. Spinoza, B.: 1949, in J. Gutmann (ed.), *Ethics*, Hafner, New York, Pars V, Praefatio.

SECTION I

HISTORICAL FOUNDATIONS OF MODERN NEUROLOGY

WILLIAM F. BYNUM

VARIETIES OF CARTESIAN EXPERIENCE IN EARLY NINETEENTH CENTURY NEUROPHYSIOLOGY [1]

I

"There probably is in the whole range of science no problem the solution of which is more difficult than that of the relation of mental faculties to particular parts of the nervous system" ([6], p. 251). When the English surgeon, Benjamin Collins Brodie, wrote those words in 1854, the systematic attempt of the German physician and anatomist, Franz Joseph Gall (1758-1828), to relate discrete mental faculties to particular parts of the brain had largely been discredited. Gall first elaborated his system in the 1790's, though it was not called phrenology until 1815 [25]. Although waning in influence by the time Brodie was writing almost thirty years after Gall's death, Gall's work represented a serious attempt to elucidate the problem which Brodie saw as soluble but as yet unsolved. I should like today to enquire into one important origin of the confidence shared by Gall and Brodie: that mental functions *are* located in different parts of the nervous system. That confidence stemmed in part from their mutual belief that the universe we inhabit is designed. One manifestation of this design is the fact that, in living organisms, structures and functions are perfectly adapted. This perfect coordination of structure and function could be found with equal ease in higher organisms such as man and in lower organisms such as insects. It could be found equally in the nervous system and the digestive system.

The demonstration of this coordination was a primary activity of the natural theologian, a species of creature which thrived in Gall's day, particularly in the Britain of William Paley and the authors of the Bridgewater Treatises. Although Gall would probably not have appreciated being classed as one of them, phrenology rested on the natural theological belief in an orderly and coherent universe. I have called this belief theological, because among Gall's contemporaries it usually was. However, it was by no

S. F. Spicker and H. T. Engelhardt, Jr. (eds.), Philosophical Dimensions of the Neuro-Medical Sciences, 15–33.

means always Christian, and the ultimate object of the natural theological demonstration was not necessarily the God of Christianity. It could be Aristotle's Prime Mover or simply Nature, written with a capital *N*. Gall himself was a deist, a fact which accords well with the fear of not a few nineteenth-century churchmen that natural theology could reveal only some impersonal creator of a clockwork universe. Even Paley himself admitted that design could best be shown through the mechanical contrivances exhibited by plants and animals.[2]

Natural theology posited the adaptation of structure to function, but it could not specify the precise way in which structure and function are related. However, I know of no natural theologian who believed that the relationship was capable of simple reduction, i.e., that function could be reduced to the mere physical and chemical properties of structure. Nevertheless, the high premium which was placed on *comparative* anatomy implied that the nature of this structure/function relationship remained constant from species to species. Let me spell this out. If structure and function are related, knowledge of function should be enhanced by more precise knowledge of structure. In particular, anatomists pursued comparative anatomical studies in the belief that (for example) information about the structure of the digestive system in a variety of species would help elucidate the particular way in which digestion was effected in each individual species. I have elsewhere called the pursuit of anatomy in the service of physiology the anatomical method, and I need do no more than remind you that the physiological or functional information yielded by anatomical knowledge was of a general kind. Detailed physiological understanding came about through the separation of anatomy and physiology and the development of an experimental physiology – events which occurred during the nineteenth century through the work of Magendie, Flourens, and Bernard in France, and Ludwig, Helmholtz, and many others in Germany.[3]

In placing Gall in this anatomical – as opposed to the experimental – tradition, I am suggesting that his work rested on the method of elucidating function through structure and observation which was widely shared by his contemporaries. I am suggesting in particular that both phrenology and comparative neuroanatomy – and by inference comparative neurophysiology – as pursued in 1800, explicitly relied on a continuity of neurological function between men and animals. Implicitly, it also presupposed a psychological continuity. The nature of this interspecific psychological continuity can be seen more clearly in Gall's work than in that of many of his contemporaries, and it made Gall a figure of controversy from the very beginning. Forced to leave Vienna because of the alleged materialistic implications of his teachings and finally denied burial on consecrated ground, Gall himself believed that he had found a new proof of the existence of God through his discovery of a specific organ or faculty of religion in man's cerebral hemispheres. Denounced by some as a vile trickster, he was hailed by others as the greatest scientist since Newton, indeed, as the Newton of the mind [2, 32].

In fact, Gall was neither a Newton nor a charlatan. He was an outstanding neuroanatomist who described a new way of dissecting the brain and identified many new structures within the central nervous system, for example, the nuclei of the first eight cranial nerves. Further, he clearly distinguished between the white and the grey matter in the nervous system and rescued the grey matter of the cerebellar and cerebral cortices from the comparatively minor roles which anatomists and physiologists had hitherto assigned it. He was also the founder of the phrenological movement, which in the hands of disciples like J.C. Spurzheim and George Combe was an important phenomenon during the first half of the nineteenth century. Phrenology has at last been rescued from a historiography which dismissed it as a pseudo-science of little consequence. It is now recognized that the phrenological movement was a significant manifestation of the popularization of science, a coherent vehicle of social and institutional reform, and a concrete embodiment of widely-shared nineteenth-century beliefs about human nature [1, 9, 11, 12, 27].

These aspects of Gall's anatomical work and of the movement which he founded are beyond dispute. But I have still to justify my choice of subject for a conference which is supposed to explore "the philosophical dimensions of the neuro-medical sciences." I have yet to provide any reason why there was considerable contemporary interest among philosophers in phrenology: why, for instance, the leading Scottish philosopher, Sir William Hamilton, was driven by phrenology into a brief excursion into experimental physiology; why the outstanding British phrenologist, George Combe, applied for the chair of logic at the University of Edinburgh, or why he wrote a work on moral philosophy; why the Scottish philosopher, Alexander Bain, left an extended account of phrenology; why the French physiologist, Pierre Flourens, dedicated his 1842 assault on Gall's theories "to the memory of Descartes"; or why philosophers such as Auguste Comte and Herbert Spencer incorporated sizeable chunks of phrenological theory into their respective philosophies [3, 17, 19, pp. 318ff. 35, pp. 113ff.].

The source of this philosophical dimension to the debates occasioned by phrenology can be found in Gall's exposition of a physiological psychology. Although his neuroanatomical discoveries can stand alone, he actually pursued those anatomical investigations within a wider theoretical context defined by his beliefs about the localization of psychological functions and the nature of human personality. He made strong claims about the brain dependence of mental processes, claims which were interpreted by some of his critics as materialistic and which at the least ran counter to the assumptions of traditional introspective analysis of mind. In developing these theories of psychological localization, Gall thus challenged certain presuppositions about the unitary, indivisible nature of the human mind. He believed that the mind was composed of a congeries of distinct faculties, each faculty having its own organ in the cerebral or cerebellar cortices. Modern theories of cerebral localization are generally seen ultimately to derive from this so-called 'organology' of Gall. Thus, Robert Young's recent historical study of "cerebral localization and its biological context" begins with Gall's work and ends with the experimental demonstrations by David Ferrier in the 1870's of the motor centers in the cortex. Young concluded that Gall contributed a conceptual framework which drew attention to the importance of the grey matter of cortices and which could handle the complicated issues involved in relating mind to brain. These contributions were recognized by writers in the second half of the nineteenth century who sometimes spoke of the work of Ferrier and others as constituting a 'new phrenology' ([37], esp. chs. 1, 2, and 8).

It should be noted, however, that Ferrier's experimental discovery of the motor strips in the cortex had strong historical connections with the other major input to the nineteenth-century concern about the localization of functions within the nervous system. This derived from the work of Charles Bell, François Magendie and others on the separation of spinal nerves into sensory and motor roots. Their work postdates phrenology. Whereas Gall was lecturing on phrenology as early as 1795, it was not until 1811 that Charles Bell correctly identified motor activity with the anterior roots of the spinal nerves. However, Bell also believed that these same roots carried sensation as well, suggesting that the functions of the posterior roots had to do with "the secret operations of the bodily frame," roughly equivalent to what would today be called autonomic. It was only in 1822 that Magendie definitively established through experiment that the anterior roots are motor, the posterior sensory.[4] In the 1830's and 40's Marshall Hall, Johannes Müller and others developed the modern concept of the reflex spinal arc, a concept which was applied to higher levels of neurological function by men such as Thomas Laycock and William B. Carpenter [37, 8, 14, 24, 28].

This tradition of sensory-motor physiology was crucial in establishing the power of experimentation in the elucidation of function within the nervous system. To the extent that it dealt with the physiological categories of sensation and motion instead of the psychological categories with which Gall was concerned, it sounded the knell for phrenology. More immediately, however, experimental physiology pointed towards an identifiable instance of specific functions associated with particular neurological structures. As the English anatomist R.D. Grainger recalled in 1837,

> Since the splendid discovery of Bell [Magendie would have been more precise], it has been justly admitted that where there is diversity of function, there must be diversity of structure; or, in other words, that each primitive fibre of the nervous system, has the property of carrying an impression in one direction only, – the fibre of sensation, from the skin to the brain, – the fibre of motion, from the brain to the muscle ([20], pp. 14-15).

The fact that Grainger here connected the operations of the sensory and motor fibers directly with the brain specifically recalled Bell's work. Notice, however, that Grainger accepted that sensory-motor physiology furnished *prima facie* evidence for some more general account of localization within the brain; supported, in fact, the proposition that "where there is diversity of function, there *must be* diversity of structure." That 'must be' represents a strong assertion which was one of the underlying assumptions not only of phrenology but, as we have seen, of most physiology in the early nineteenth century. Grainger's comments, however, avoid explicit reference to any possible relationship between a *psychological* function, on the one hand, and a *physiological* one on the other. And the actual discoveries of sensory-motor physiology did not provide any obvious answer. In fact, Ferrier's account of cerebral localization had relatively little in common with the psychological categories which Gall had placed in the cortex more than half a century earlier. However, Paul Broca's work in the 1860's on the localization of the center of speech can be placed more precisely in the direct phrenological stream.[5]

On the matter of the kind of localization found within the nervous system, then, spinal nerve root physiology offered no support for phrenology. Neither Bell nor Magendie accepted phrenology, and Flourens in particular used the experimental techniques so succesful in the case of the spinal nerves to substantiate his damning critique of Gall in 1842. This same diffidence to move from brain anatomy and physiology to psychology had also characterized one of the earliest official judgments on the work of Gall and Spurzheim, the report which Cuvier, Pinel, and other distinguished scientists prepared for the French Academy of Sciences in 1808. This group refused to comment on Gall's psychological doctrines, since psychology, they wrote, "ultimately depends upon observations relative to the moral and intellectual disposition of individuals, which certainly are not within the sphere of any academy of sciences" [33]. In other words, Gall's psychological theories were not the proper business of the scientist *qua* scientist, an opinion endorsed by the Scottish philosopher, Sir William Hamilton, who proposed to test phrenology by the same kind of cerebral ablation experiments which Flourens had pioneered in France. These experiments – albeit rather crude – convinced Hamilton "that no assistance is afforded to Mental Philosophy by the examination of the Nervous System, and that the doctrine, or doctrines, which found upon the supposed parallelism of brain and mind, are, as far as observation extends, wholly groundless."[6]

Embodied in this brief introduction to the question of neurological localization in the early decades of the nineteenth century are three different views of Gall's endeavour. There was Gall's own view that psychological faculties are localized in various parts of the brain and that it is possible to discover these sites through dissection, measurement, and observation. There was the alternative view that dissection and experimentation could yield direct evidence of the localization of functions within the nervous system. While this experimental physiology did uncover the localized functions of the spinal nerve roots, the direct extension of experimental methods to the study of the brain itself was not propitious for the specific, localized, psychological categories which Gall had proposed. In fact, William Hamilton used these experimental techniques to underpin a third view of phrenology, namely, that there being no evidence for psycho-physical parallelism in the brain, nothing is to be learned about the functions of the mind from studying the brain. Superficially, at least, Hamilton's opinion was shared by many of his contemporaries, including anatomists and physiologists. When George Combe attended the anatomical lectures of the eminent Scottish anatomist, John Barclay, in the 1810's, he watched patiently for four hours while Barclay methodically dissected a human brain. Combe, who had already abandoned Locke and Dugald Stewart as muddled and incomprehensible, was dismayed when Barclay, the dissection over, announced that all he had been communicating "amounted to nothing more than a display of parts of the brain in the order of an arbitrary dissection, and that in simple truth nothing was known concerning the relation of the structure which he had exhibited and the functions of the mind" ([19], I., p. 93).

In the committee of the Academy of Sciences, in William Hamilton, and in John Barclay, we find three instances in which knowledge of brain anatomy was explicitly disclaimed as one avenue for insight into the behavioral characteristics of men and animals. Was neuroanatomy thus looked upon as the rather arbitrary endeavour that Barclay made it out to be? I would suggest that it was decidedly not. Embedded in the anatomical method, referred to earlier and held by most early nineteenth-century life scientists, were assumptions similar to the three fundamental principles on which Gall based his system of phrenology. In particular, comparative neuroanatomists had difficulty consistently handling a concept of mind which was not equally applicable to men and animals. Indeed, the assumptions on which Flourens and Hamilton based their experimental proofs of the unitary character of mind presupposed just such a psychological continuity. And while psychological continuity might not necessitate psychological localization, localization was implied by the commonplace comparative observation that the brains of different species varied in the complexity and arrangement of the component parts of their nervous systems. From this perspective, Gall is perhaps more of a traditional figure than an innovator. I should like now to examine this claim by looking at the assumptions on which phrenology was based.

II

I have said that I should like to see Gall as a traditional figure. This is neither to deny his historical importance nor to assert that his ideas were all picked up from his predecessors. He was, of course, influenced by some of the authors he had read. Erna Lesky has recently drawn attention to certain features in the works of the French philosophical biologist, Charles Bonnet, and the German philosopher, J.G. Herder, which Gall incorporated into his own systematic thought [23]. Gall himself acknowledged his debt to these men. His own description of the origins of his discovery of his system also reveals another major intellectual debt. Gall first hit upon the idea that psychological traits could be located by noting that his schoolfellows with the best memories had unusually prominent eyes ([18], pp. 2-3). Memory was always fixed by phrenologists in the frontal region of the brain. But it was the observation of skull shape and facial features rather than brain shape which first led Gall to attempt the correlation of anatomy and psychology, a fact which alerts us to the historical affinities which phrenology had with physiognomy, the art of reading character from the form of the face. As a coherent genre, physiognomy reaches back to antiquity, though its major modern popularizer was the eighteenth-century Swiss clergyman, J.C. Lavater. So important was this tradition for Gall that he frequently called his system a physiognomical one.[7]

The belief that character and form relate in some intelligible fashion is above all a belief about meaningful order in the world we inhabit. Physiognomy is still part of our everyday lives, one of those frequently unstated assumptions which enter into our appreciation of art and literature. Physiognomy was perhaps a more conspicuous feature of nineteenth-century life, yet we still respond and need not be told when reading *Nicholas Nickleby* whether Wackford Squeers is a good or bad character when Charles Dickens informs us that

> [Squeers] had but one eye, and the popular prejudice runs in favour of two. The eye he had was unquestionably useful, but decidedly not ornamental: being of a greenish grey, and in shape resembling the fanlight of a street door. The blank side of his face was much wrinkled and puckered up... . His hair was very flat and shiny, save at the ends, where it was brushed stiffly up from a low protruding forehead ([13], ch. 4).

Surely no hero of Dickens would have "a low protruding forehead," though I know of no *a priori* reason why such an attribute might not exist in the best of fellows.

Physiognomy postulated the existence of a meaningful relationship between an assessed function (character) and an observed structure (form). In linking a structure to a function, physiognomy was a relational endeavour with conceptual affinities with the more familiar practice of what Albrecht von Haller called 'animated anatomy,' and which we have already identified as the 'anatomical tradition'. I pointed out that Grainger's belief that anatomical structure and physiological function are such that diversity of one implies diversity of the other. Like the belief that character and form are related, this faith that structure and function are congruent stemmed at its deepest level from the belief in an orderly universe. Nature is perfect, she does nothing in vain, each part of an organism has its own proper function, a functionless structure would be a blight on the creation. These sentiments are all expressed by Aristotle and Galen, and were codified and elaborated by countless subsequent authors. During the period we are concerned with, they survived explicitly in the popular genre of natural theology, but they also characterized the workaday assumptions of the ordinary anatomist *cum* physiologist. They are also an integral part of the architecture of phrenology. Phrenology's fundamental assumptions remained constant throughout the history of the movement. They were succinctly stated by George Combe as consisting of the following three 'fundamental principles':

(1) "That the brain is the organ of the mind";

(2) "That the brain is an aggregate of several parts, each subserving a distinct mental faculty";

(3) "That the size of the cerebral organ is, *caeteris paribus*, an index of power or energy of function" ([19], I., p. 204).

I should like to examine each of these principles in turn, setting it within the context of early nineteenth-century thought.

1. *The brain is the organ of the mind.* In the usual phrenological exposition, this was not an identity theory; indeed, in the way it was phrased it suggested that the brain and the mind are different. That the brain is the organ of the mind was supported by a variety of evidences drawn from medicine and from ordinary life: by the fact that consciousness can be destroyed by a blow to the head or by an attack of apoplexy; that it can be disturbed by disease or deranged by insanity; that the mind seems to have a life history of its own, which parallels the developmental history of the brain, from the soft pliable brain of the infant, through its firm, relatively distinct features in the adult, to its atrophied state in the senile. Both phrenologists and non-phrenologists granted the impossibility of mental development in the acephalic infant, a fetal abnormality which attracted a good deal of careful study during the period (e.g., [22]). While it

is true that a few phrenologists interpreted this proposition materialistically and asserted that brain processes *cause* mental phenomena, most phrenologists followed Gall in remaining ambiguous, merely asserting that the exercise of mental faculties depends on the material conditions of the brain but are not necessarily the product of them.[8] In fact, the standard phrenological position was rarely much stronger than that of one of their critics, Charles Bell, who wrote in 1824:

> If... it should be found that the mind is dependent on the frame of the body, the discouvery ought not to be considered as humiliating.... It is a fundamental law of our nature that the mind shall be subject to the operation of the body, and have its powers developed through its influence.... Since we are dwellers in a material world, it is necessary that the spirit should be given up to the influence of a material and organized body, without which it could neither feel, nor re-act, nor manifest itself in any way ([4], pp. 15-16).

Bell thus believed that this body-dependence of mind was neither materialistic nor inimical to the tenets of natural theology, which he espoused as the author of one of the Bridgewater Treatises and as an editor of a revised edition of Paley's *Natural Theology* [5, 7]. Likewise, Bell's work on the localized functions of the roots of the spinal nerves was set in a larger conceptual framework which posited that the cerebrum is the specific organ of the higher mental functions.[9] Bell's criticisms of phrenology were aimed at its details, not at its basic principles with which he was in essential agreement.

2. *The brain is an aggregate of several parts, each subserving a distinct mental faculty*. No anatomist of Gall's day would have quarrelled with the first part of this proposition. The cerebrum obviously differs from the cerebellum, and the fornix, ventricles, corpus callosum, quadrigeminal bodies, the pineal body, etc. were all recognized as structures sufficiently distinct to warrant separate names. They were different *parts* of the nervous system, to use the traditional term of Galen's treatise on anatomy and physiology, *De usu partium*. As we have seen, it was a corollary of the notion that nature does nothing in vain, that each part has its function or functions. To be sure, in 1800 there was very little precise information about the functions of any part of the brain, but there was a rich history of assigning various uses to the brain or parts of the brain, ranging from making phlegm or cooling the blood to the rather more exalted activities of imagining, reasoning, and remembering. These latter activities or faculties

had been associated with the ventricles in antiquity and the Middle Ages, and one of Gall's contemporaries, Samuel Thomas Soemmerring, proposed on anatomical grounds in 1796 that the ventricular walls contain the convergence of the sensory nerves, thus forming the *sensorium commune* or common sense ([29], pp. 10-11). Sensorium commune had long been regarded both anatomically and philosophically as the seat of the soul, since it was there that sensations are perceived. Soemmerring was only one of a number of late eighteenth-century anatomists who believed he had located it (see [15], [21], [30]). On more general anatomical grounds, Thomas Willis in the seventeenth century elaborated a comprehensive scheme of both neurological and psychological localization. He had located intelligence in the cerebrum, involuntary movement in the cerebellum, sense perception in the corpus striatum, imagination in the corpus callosum, and instinctive behavior in the 'mid brain'. Willis' system of localization was presumably no more nor less materialistic than Descartes' own comments on the role of the pineal body, though when Willis shrewdly located man's intellectual faculty in the cerebrum on the grounds that this part of the brain was proportionately larger and more complex in man than in other animals, he implied a kind of psychological continuity not entirely compatible with his stated belief in man's unique, rational soul and its functions.[10] In this context, the usual caveat issued by Willis, Soemmerring, Bell, and Gall, that the nature of the union of body and mind will always be obscure, applied with equal force to the case of animals.

Seen from this perspective, Gall's theories acquire a more familiar character. Certainly he multiplied the number of psychological faculties which were presumed to have discrete anatomical localization, and he specified rather precisely where he thought they were located. But his postulated psychological categories were by no means all new. Many phrenological faculties can be found discussed by the eighteenth-century Scottish common sense philosophers such as Thomas Reid and Dugald Stewart.[11] Phrenologists readily acknowledged this fact. One of them pointed out that the philosophical works of Reid's friend, Lord Kames, identify twenty out of the thirty-odd faculties ultimately postulated by phrenologists ([10], p. 147). The quaint and distinctive vocabulary with which they named their faculties (philoprogenitiveness, amativeness, etc.) should not obscure phrenology's conceptual filiations with previous faculty psychologies. And both common sense philosophy and phrenology represented attempts to bring the study of the mind into closer accord with the observed behavioral complexities of everyday life.

At the same time, Gall sought to distance himself from his philosophical predecessors by eschewing their subjective methods of introspection. He insisted that his own system was based on the objective observations of behavior, talents, inclinations, and activities of men and animals, combined with the detailed correlation of these with cranial form. He was always in search of the extremes of humanity – great talent in one sphere, excessive vice in another – since he believed that these extremes would define his categories more precisely (cf. [37], esp. pp. 18-19 and 23 ff.). The fact that these faculties could be correlated with cranial (as opposed to brain) form gave phrenology a wide practical orientation, though as a system of psychology, phrenology was not of necessity bound to Gall's contention that cranial form accurately reflects the underlying cortical surfaces. It is, however, an historically significant facet of the tenuity of Gall's organology that he first derived his idea from the cranium rather than the brain. I doubt if anyone would ever have approximated a theory as detailed as phrenology by starting with the brain surfaces.

Craniology made phrenology into a living psychology and provided the materials which in theory could have tested it. Accurate measurements of crania correlated with detailed assessments of psychological capacities should have been definitive. In practice, of course, it was impossible to falsify phrenology through craniometry since phrenology was flexible enough to encompass apparent anomalies, and since the faculties it posited could be neither defined nor measured with sufficient precision. On the other hand, competent practical phrenologists impressed a number of intelligent individuals with their capacity to give character assessments which rang true (e.g. [36]). In fact, such assessments provided a major source of recruitment to the movement, though critics remained generally unconvinced by interpretative craniology. However, it was possible (though not common) for overly zealous phrenologists to explain why critics refused to accept their science by reference to the cranial deficiencies of these critics. Phrenology might be said to share with Marxism and psychoanalysis the characteristic which Karl Popper has called 'reinforced dogmatism' – the capacity to explain away critics' lack of assent by reference to the principles of the system under debate. As one phrenological writer put it, "If phrenology be *true*, no man can possibly oppose it who is not either uninformed concerning it, limited in intellect, so as to be unable to comprehend it, or destitute of honesty to admit the conviction which he feels."[12]

Phrenologists thus proposed that their science be tested by inspection rather than introspection, a difference which underscores the new dimension which Gall attempted to give to the faculty concept. Cranial inspection documented numerous contours believed to reflect the cortical surfaces just beneath. In contrast, introspection revealed an undivided *moi* or ego, and this unity of consciousness had been taken as the equivalent of the indivisible mind. Mind could manifest various faculties or powers, but mind in itself was not divisible because not material. Pierre Flourens stressed that the chief flaw inherent in phrenology was its failure to recognize this unity of the conscious mind. Gall and his followers practised bad psychology, bad physiology, and bad philosophy, and Flourens wanted to replace phrenology with a philosophical psychology and an experimental physiology. He deemed experimental physiology capable of documenting the essential Cartesian message about the fundamental distinctions between unified mind and divided body ([16], esp. p. 236ff.).

Now, it is easy to contrast Flourens and Gall, both as individuals and as interpreters of nature. Flourens was a powerful member of the French scientific establishment, an insider who succeeded Cuvier as perpetual secretary of the Academy of Sciences and who was elected to the French Academy in favor of Victor Hugo. Gall was a German exile in France who eventually became a French citizen but never held any official post. Flourens believed in the unity of mind, whereas Gall held it to be a congeries of separate faculties. Flourens was an experimentalist; Gall abhorred vivisection and felt that direct experimentation could never reveal the higher functions of the brain. Flourens can be said to symbolize the establishment which never really accepted either Gall or his movement (cf. [37], ch. 2, [26]).

These dichotomies are significant in both social and conceptual terms. Yet, we should not forget that – like Gall – Flourens accepted the general premise that both neurological and psychological functions are localized. He divided the nervous system into six major units: nerves, spinal cord, medulla, quadrigeminal bodies, cerebellum, and cerebrum. Through controlled ablation experiments, he believed himself able to demonstrate the relevant function or functions of each unit from an analysis of the neurological or psychological deficit which the animal exhibited following ablation. Flourens' classical experiments on the cerebellum were published in the same year (1822) that Magendie localized motor and sensory functions to the anterior and posterior roots of the spinal nerves. Flourens established the role of the cerebellum in coordinating the activity of skeletal muscles into smooth, effective movements. Deprived of its cerebellum, the animal became clumsy and uncoordinated. Extirpation of the quadrigeminal bodies resulted in blindness, implicating these structures in sight. Removal of the cerebral hemispheres left the laboratory animal stupefied and unable to initiate any volitional act. Whereas it would swallow if food were placed in its mouth, if left alone it would remain still until it starved to death, even if food were placed nearby. It breathed, digested, excreted, withdrew from painful stimuli, but neither avoided danger nor initiated any action. On the other hand Flourens found that serially slicing away the cerebrum gradually diminished these higher functions of memory, intelligence, perception, and volition in proportion to the amount of cerebral tissue removed. This kind of experiment formed the bulwark of his theory of equipotential cerebral action, which in turn represented the anatomical equivalent to this belief in the unified mind (cf. [31]).

Much has been made of the contrast between Flourens' model of the equipotential cerebrum and Gall's insistence that the brain contains many discrete psychological faculties. It is sometimes overlooked, however, that Gall's handling of the traditional faculties of scholastic philosophy was, in practical terms, not so very different from Flourens'. For Gall did not include general mental functions like volition, intelligence, judgment and understanding in his list of individual faculties. Rather, he proposed that these functions were associated with each specific faculty. In Gall's terms, what Flourens called "general intelligence" was the cumulative effect of many specific intelligences, e.g., an intelligence for form, for colour, for calculation, etc. And although Flourens sarcastically demanded of Gall an answer to his query, "Where is general intelligence?", Gall's solution was as general as that of Flourens since, according to phrenologists, *all* of the grey matter of the cerebrum and cerebellum contributed towards intelligence.[13] On the other hand, by refusing to break intelligence down into any simpler units, Flourens was unable to provide a consistent account why individuals do in fact differ. The higher psychological functions were, in his opinion, metaphysical attributes. Yet he tried to meet Gall on the common ground of cerebral anatomy and physiology. The difficulties of doing that can best be seen in conjunction with Gall's third fundamental proposition, to which we now briefly turn.

3. *That the size of the cerebral organ is, caeteris paribus, an index of power or energy of function.* Gall, of course, integrated this proposition into his organology. It formed the rationale for correlating specific brain or skull contours with specific personality traits or psychological characteristics. In practice, the 'caeteris paribus' clause covered a multitude of sins, since integrated behavior demanded a proper balance of all the faculties and not just a prominence or deficiency of any one. However, the details of phrenological practice need not concern us. More significant is the general way in which anatomists had always used the proposition that – where the brain is concerned – big is beautiful. Was it not significant that men possess larger brains than cats or dogs? And in natural theological terms, that significance lay in the fact that man's refined, complex, and capacious brain was a suitable house for his rational soul. It was at once manifestation and evidence of design.

At the same time, comparative anatomists pointed out that the whale and elephant have larger brains than men, and naturalists knew that the small-brained raven is more intelligent than the larger brained crocodile. What do we make of this? Late in the eighteenth century Blumenbach and Soemmerring suggested that brain/body ratio would give a fairer assessment than absolute brain size. But on this basis sparrows did better than men. Accordingly, Soemmerring and other anatomists subsequently modified the parameters and measured the ratio between brain mass and the mass of the nerves deriving from the brain. This would correct for the portion of the brain concerned with the so-called 'animal functions' and leave a truer basis for comparing intelligence.[14]

Significantly, 'intelligence' was the parameter which anatomists believed could be correlated with some structural property of the nervous system. The parameter was ill-defined, to be sure, and was frequently based on anecdote and legend about the intellectual capacities of certain animals or species. But the human species was almost invariably included in any attempt to correlate brain properties with intelligence. And Flourens sought to demonstrate the unity of the *human* mind by successively slicing away parts of the cerebrum of birds, dogs, cats, and other laboratory animals. If, as he believed, he had proved that the vertebrate cerebrum works as a unit, he demonstrated the unity of human minds only assuming a unity of animal minds. And at this point, Flourens' self-conscious espousal of Cartesian dualism broke down, and he was left with Descartes' legacy rather than the sharply-defined universe of Descartes himself. That legacy, as Vartanian and others have documented, was the choice between animal souls and human machines. By implication Flourens chose the former, while La Mettrie had in the middle of the eighteenth century notoriously opted for the latter (see esp. [34]). Although Flourens returned to Descartes for inspiration, he was closer to Gall than he realized, for both his work and Gall's presupposed a feature common to the notion of the animal soul and to the suggestion that men are machines. That feature is a doctrine of psychological continuity – not in the sense that all human psychological attributes can be found in animals, but in the sense that human minds and animal minds are equally dependent on neurological development. Gall attributed man's uniqueness to special cerebral organs whereas Flourens spoke only of general psychological functions common to men and animals. Some form of psychological continuity was needed to underpin the search for parameters which would correlate brain and intelligence. Indeed, the fact that Flourens and Gall each placed such a high functional premium on the cerebrum buttresses this point, for the cerebrum is that part of the nervous system where interspecific variations are most striking. In nineteenth-century terms, the cerebrum alone could sustain a theory of psychological continuity grounded on anatomical structure. There were thus good reasons for locating the higher psychological functions there. After all, man's large, complex cerebrum must be there for some purpose. The argument from design quaranteed that. And at this point, Gall and Flourens were in assent with Bell, Paley, and other natural theologians.

In thus yoking Gall and Flourens together, I hope I have not distorted the important conceptual issues separating them. Their respective philosophies of mind were divergent and in some ways incompatible. But both their philosophies of mind rested on the ascription of minds to animals. It is this biological dimension of the phrenological debates which I have emphasized today.

University College London,
London, England

NOTES

[1] I am grateful to Professor Erwin H. Ackerknecht, Dr. M.J. Bartholomew, and Mr. B.J. Norton for their helpful comments on an earlier draft of this paper.

[2] Paley, W.: 1823, *Natural Theology*, in *Works*, Vol. 3, London, p. 52: "the *mechanical* parts of our fame,... although constituting probably the coarsest portions of nature's workmanship, are the most proper to be alleged as proofs and specimens of design." On the dangers of natural theology, cf. Newman, J.H.: 1960, in M.I. Svaglic (ed.), *The Idea of a University*, Holt, Rinehart and Winston, San Francisco, p. 340: "I confess, in spite of whatever may be said in its favour, I have ever viewed [natural theology] with the greatest suspicion." The God of the natural theologian may be "not very different from the God of the Pantheist." (p. 342).

[3] This process is surveyed by Joseph Schiller: 1968, 'Physiology's Struggle for Independence in the First Half of the Nineteenth Century', *History of Science* 7, 64-89. I have examined some of these themes for the seventeenth century in 'The Anatomical Method, Natural Theology, and the Functions of the Brain', *Isis* **64**, (1973), 445-468.

[4] This literature has recently been reprinted with an introduction by Cranefield, P.F.: 1974, *The Way In and The Way Out*, Futura Pub. Co., Mount Kisco, New York.

[5] Although as Young has noted, Ferrier later attempted to integrate his researches on cerebral localization into a psychophysiological framework ([37], p. 243).

[6] Hamilton, W.: 1865, *Lectures on Metaphysics and Logic*, in H.L. Mansel and J. Veitch (eds.), 4 vols., I., Edinburgh, p. 264n. The literature of Hamilton's ongoing debate with George Combe was printed in vols. 4 and 5 of *The Phrenological Journal*. Hamilton's physiological "researches" are described in Veitch ([35], p. 117).

[7] Gall, however, sought to distance himself from his physiognomical predecessors. Nevertheless, in its popular form, phrenology continued to be practised within the context of physiognomy. Cf., e.g., Spurzheim, J.C.: 1815, *The Physiognomical System of Drs. Gall and Spurzheim*, Baldwin, Craddock, and Joy, London; and Wells, S.R.: 1896, *New Physiognomy or Signs of Character*, Fowler and Wells, New York. Lavater's works also continued to be published throughout the nineteenth century.

[8] Cf. Gall (18, I., p. 189): "Quand je dis que l'exercice de nos facultés morales et intellectuelles dépend de conditions matérielles, je n'entends pas que nos facultés soient un *produit* de l'organisation; ce serait confondre les conditions avec les causes efficientes." For a materialistic statement of phrenology, cf. Engledue, W.C.: 1842, *Cerebral Physiology and Materialism*, London.

[9] Bell, C.: 1811, *Idea of a New Anatomy of the Brain*, Strahan and Preston, London, reprinted in Cranefield (n. 4, p. 27): "The Cerebrum I consider as the grand organ by which the mind is united to the body."

[10] For a lengthier discussion of these issues, cf. Bynum (n. 3).

[11] On faculty psychologies, cf. Boring, E.G.: 1950, *A History of Experimental Psychology*, 2nd ed., Appleton-Century-Crofts, New York, ch. 11. Spoerl, H.D.: 1935-6, 'Faculties Versus Traits: Gall's Solution', *Character and Personality* 4, 216-231, has argued that Gall was not influenced by Scottish faculty psychologies. Cf. also Young, R.M.: 1966, 'Scholarship and the History of the Behavioral Sciences', *History of Science* 5, 1-51, esp. pp. 14-15. However, my point is not concerned so much with the origin of Gall's ideas as in the consonance between Gall and his predecessors.

[12] [Anon.]: 1823-24, 'Signs of the Times', *Phrenological Journal* I, 316-319, p. 317. Geoffrey Cantor pointed out this reference to me. On 'Reinforced Dogmatism', cf. Popper, K.: 1966, *The Open Society and its Enemies*, 5th ed., 2 vols., Routledge and Kegan Paul, London, e.g., pp. 40, 215-216.

[13] For Flourens' criticisms, cf. Flourens ([17], esp. ch. 2).

[14] These issues are discussed by Spurzheim (n. 7, pp. 190-207), and Blumenbach, J.F.: 1827, *Manual of Comparative Anatomy* (transl. by W. Lawrence), 2nd ed., Simpkin and Marshall, London, pp. 203-257.

BIBLIOGRAPHY

1. Ackerknecht, E.H.: 1958, 'Contributions of Gall and the Phrenologists to Knowledge of Brain Function', in F.N.L. Poynter (ed.), *The Brain and its Functions*, Oxford.
2. Ackerknecht, E.H. and Vallois, H.: 1956, *Franz Joseph Gall, Inventor of Phrenology and His Collection*, Madison, Wisc.
3. Bain, A.: 1861, *On the Study of Character, Including an Estimate of Phrenology*, London.
4. Bell, C.: 1824, *Essays on the Anatomy and Philosophy of Expression*, 2nd ed., London.
5. Bell, C.: 1833, *The Hand, its Mechanism and Vital Endowments as Evincing Design*, London.
6. Brodie, B.C.: 1865, *Psychological Inquiries*, in C. Hawkins (ed.), *Works*, 3 vols., I., London.
7. Brougham, H. and Bell, C. (eds.): 1836, *Paley's Natural Theology*, 2 vols., London.
8. Canguilhem, G.: 1955, *La formation du concept de réflexe aux XVIIe et XVIIIe siècles*, Presses universitaires de France, Paris.
9. Cantor, G. and Shaping, S.: *Annals of Science* **32**, 195-256, on phrenology in early nineteenth-century Edinburgh.
10. Capen, N.: 1881, *Reminiscences of Dr. Spurzheim and George Combe*, Boston.
11. Davies, J.D.: 1955, *Phrenology, Fad, and Science*, New Haven, Conn.
12. De Giustino, D.: 1975, *Conquest of Mind, Phrenology and Victorian Social Thought*, London.
13. Dickens, C.: *Nicholas Nickleby*.
14. Fearing, F.: 1970, *Reflex Action*, MIT Press, Cambridge, Mass., chs. 9 and 10.
15. Figlio, K.M.: 1975, 'Theories of Perception and the Physiology of Mind in the Late Eighteenth Century', *History of Science* **13**, 177-212.
16. Flourens, P.: 1824, *Recherches Expérimentales sur les Propriétés et les Fonctions du Système Nerveux*, Paris.
17. Flourens, P.: 1846, *Phrenology Examined*, (transl. by C. Meigs), Philadelphia.
18. Gall, F.J.: 1822-1825, *Sur les Fonctions du Cerveau*, 6 vols., I.
19. Gibbon, C.: 1878, *The Life of George Combe*, 2 vols., I., London.
20. Grainger, R.D.: 1837, *Observations on the Structure and Functions of the Spinal Cord*, London.

21. Keele, K.D.: 1957, *Anatomies of Pain*, Charles C. Thomas, Springfield, Ill., ch. V.
22. Lawrence, W.: 1814, 'An Account of a Child Born Without a Brain', *Medico-Chirurgical Transactions* 5, 165-224.
23. Lesky, E.: 1970, 'Structure and Function in Gall', *Bull. Hist. Med.* **44**, 297-314.
24. Liddell, E.G.T.: 1960, *The Discovery of Reflexes*, Oxford Univ. Press, London, ch. 3.
25. Noel, P.C. and E.T. Carlson: 1970, 'Origins of the Word "Phrenology" ', *Amer. J. Psychiat.* **127**, 694-697.
26. Olmstead, J.M.D.: 1953, 'Pierre Flourens', in E.A. Underwood (ed.), *Science, Medicine, and History*, 2 vols., Oxford Univ. Press, London, pp. 290-302.
27. Parssinen, T.M.: 1974, 'Popular Science and Society: The Phrenology Movement in Early Victorian Britain', *J. Social History* 8, 1-20.
28. Smith, R.: 1970, 'Physiological Psychology and the Philosophy of Nature in Mid-Nineteenth Century Britain', Cambridge Univ. Ph.D. Thesis.
29. Soemmerring, S.T.: 1796, *Über das Organ der Seele,* F. Nicolovius, Köningsberg.
30. Soury, J.: 1899, *Le Système Nerveux Central,* 2 vols., G. Carré and C. Naud, Paris, esp. I., pp. 468 ff.
31. Swazey, J.P.: 1970, 'Action Propre and Action Commune: The Localization of Cerebral Function', *J. Hist. Biology* **3**, 213-234.
32. Temkin, O.: 1947, 'Gall and the Phrenological Movement', *Bull. Hist. Med.* **21**, 275-321.
33. Tenon, Portal, Sabatier, Pinel, and Cuvier: 1809, 'Report on a Memoir of Drs. Gall and Spurzheim...', transl. in *Edinburgh Med. and Surg. J.* 5, 36-66, p. 37.
34. Vartanian, A.: 1953, *Diderot and Descartes, A Study of Scientific Naturalism in the Enlightenment*, Princeton Univ. Press, Princeton; and 1960: *La Mettrie's L'Homme Machine, A Study in the Origins of an Idea*, Princeton Univ. Press, Princeton.
35. Veitch, J.: 1869, *Memoir of Sir William Hamilton, Bart.*, Edinburgh.
36. Wallace, A.R.: 1905, *My Life*, 2 vols., Chapman and Hall, London, Vol. I, pp. 257-262.
37. Young, R.M.: 1970, *Mind, Brain, and Adaptation in the Nineteenth Century*, Clarendon Press, Oxford.

ARTHUR BENTON

HISTORICAL DEVELOPMENT OF THE CONCEPT OF HEMISPHERIC CEREBRAL DOMINANCE

I

The term, 'hemispheric cerebral dominance,' expresses the concept that the two hemispheres of the human brain are not equivalent in function, and that each hemisphere appears to possess distinctive functional properties which are not shared by the other. The concept is in essence a restricted formulation of the more general concept of cerebral localization which states that different areas of the brain subserve different sensory, motor, or mental functions; in brief, that the human brain is not a functionally equipotential organ.

Cerebral localization is an ancient concept. As early as the first century A.D., Greek physicians and philosophers postulated an association between specific mental functions and different regions of the brain along its longitudinal axis. Perception was assigned to the anterior part of the brain, reasoning to the middle part, and memory to the posterior part. It is uncertain whether or not these conclusions had an empirical basis. Possibly they were derived from observations of patients with traumatic brain injury. Probably they arose from structural considerations and were purely speculative in nature. More often than not, the locus of these mental functions was assumed to be in the ventricles of the brain rather than in brain tissue. Thus, perception was localized in the lateral ventricles, reasoning in the third ventricle and memory in the fourth ventricle. It is likely that theological considerations provided the prime motive for this form of localization [40]. The pneumatic spaces of the ventricles seemed to be a more appropriate place for the non-corporeal soul to interact with the body than did the substance of the brain itself.

Figure 1 shows Gregor Reisch's well known and much imitated pictorial conception of ventricular localization of function which appeared in his *Margarita Philosophica*, the first edition of which was published in 1503.

S. F. Spicker and H. T. Engelhardt, Jr. (eds.), Philosophical Dimensions of the Neuro-Medical Sciences, 35–57.

Sensation, perception, and imagery take place in the anterior (lateral) ventricle which communicates with the middle (third) ventricle. The middle ventricle is the seat of cogitation, judgment, and reasoning, and it communicates with the posterior ventricle, which is the seat of memory. The connections between the ventricles form the structural basis for a dynamic process in which sensory information is received and integrated in the anterior ventricle, reflected upon in the middle ventricle, and finally placed in a memory store in the posterior ventricle.

The ventricular conception was widely adopted and held sway for a remarkably long time. It served as a basis for medical diagnosis and treatment in appropriate cases. For example, the 15th century physician, Antonio Guainerio [25], having described a patient with amnesic aphasia who was unable to produce the correct names of people familiar to him, inferred that the cause of the disorder was the accumulation of an excessive amount of phlegm in the posterior ventricle, the 'organ of memory'.

Ventricular localization of cerebral function was still accepted by a few anatomists and physicians as late as 1800. However, beginning in the 17th century, it was gradually supplanted by conceptions providing for localization of function in the substance of the brain. For example, Thomas Willis (1621-1675) placed the seat of intelligence and ideation in the corpus callosum and the seat of memory in the gray matter of the cerebral cortex. As every philosopher knows, Descartes (1596-1650) identified the pineal gland as the locus of interaction between the soul and the body, primarily because of its central position. The notion was specifically rejected by Willis, who pointed out that the pineal gland is often found to be well-developed in animals who lack intelligence and learning capacity.

It was the anatomist and phrenologist, Franz Joseph Gall (1758-1828), who made the concept of cerebral localization a central issue in scientific and medical thought. His fundamental postulate was that the human brain was not a single organ but an assemblage of organs, each of which formed the material substrate of a specific intellectual faculty or personality trait. He found no difficulty in identifying the locus in the brain of some 30 mental traits, most of which he took from the analyses of the Scottish school of faculty psychology as represented by Thomas Reid and Dugald Stewart. A phrenological conception of localization of functions in the substance of the brain is shown in Figure 2. I have selected this chart by Thoré, a follower of Gall, because it indicates the locus of the various

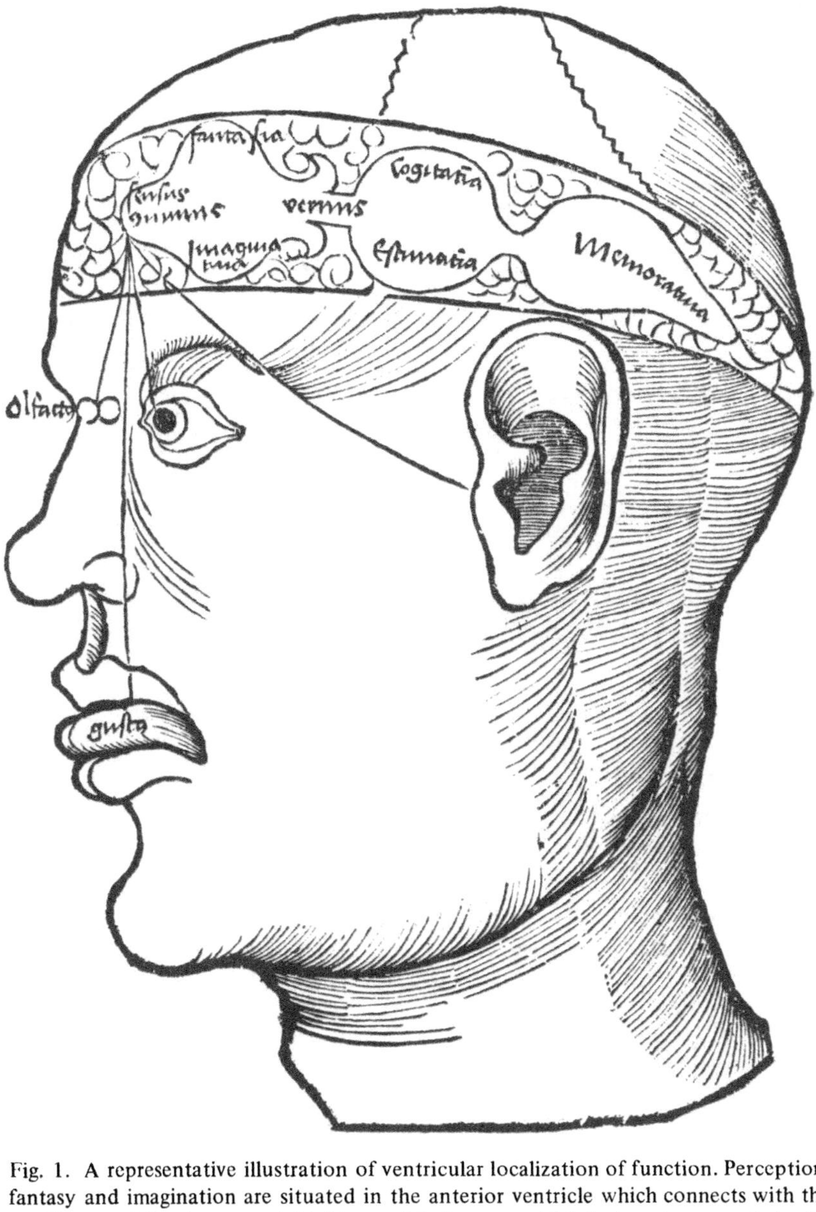

Fig. 1. A representative illustration of ventricular localization of function. Perception, fantasy and imagination are situated in the anterior ventricle which connects with the middle ventricle, the seat of thinking and judgment, which in turn is connected with the posterior ventricle, the seat of memory. The illustration is from the copy of the second edition of Gregor Reisch's *Margarita Philosophica* [41] in the Health Sciences Library of the University of Iowa.

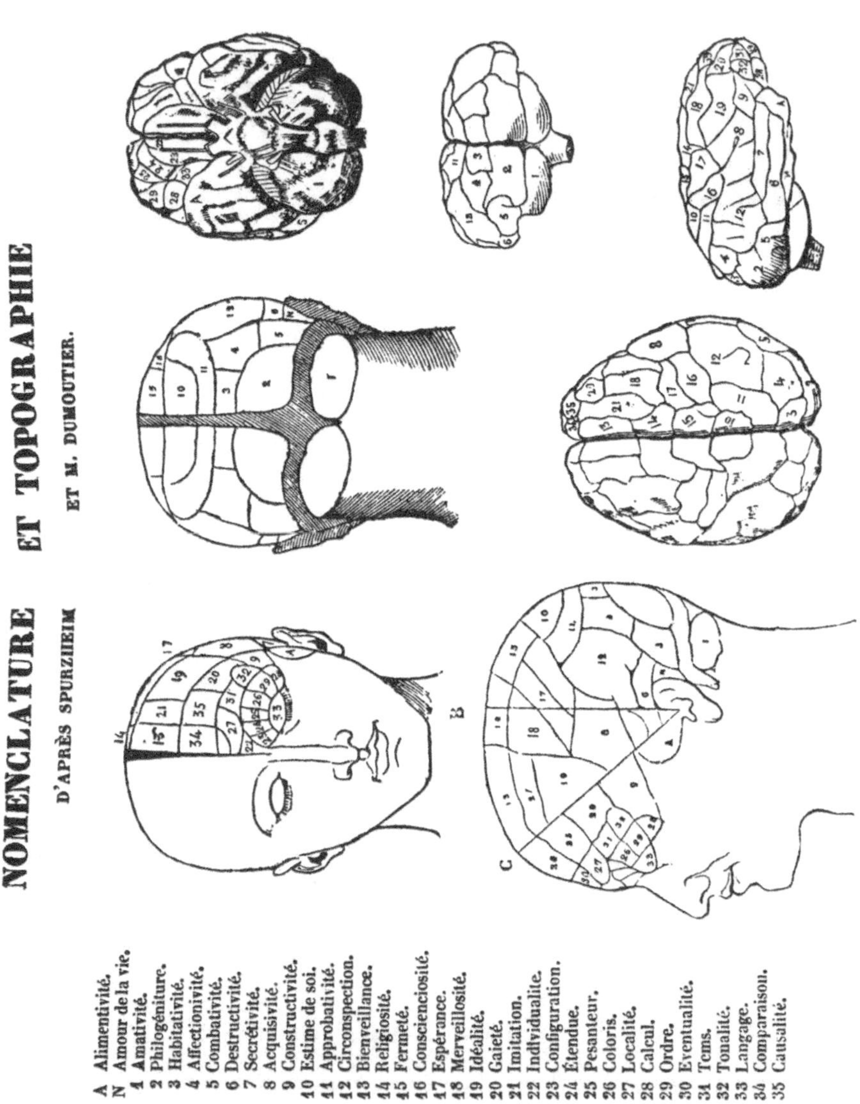

Fig. 2. A phrenological conception of localization providing for the precise cortical localization of personality traits as well as cognitive abilities [48].

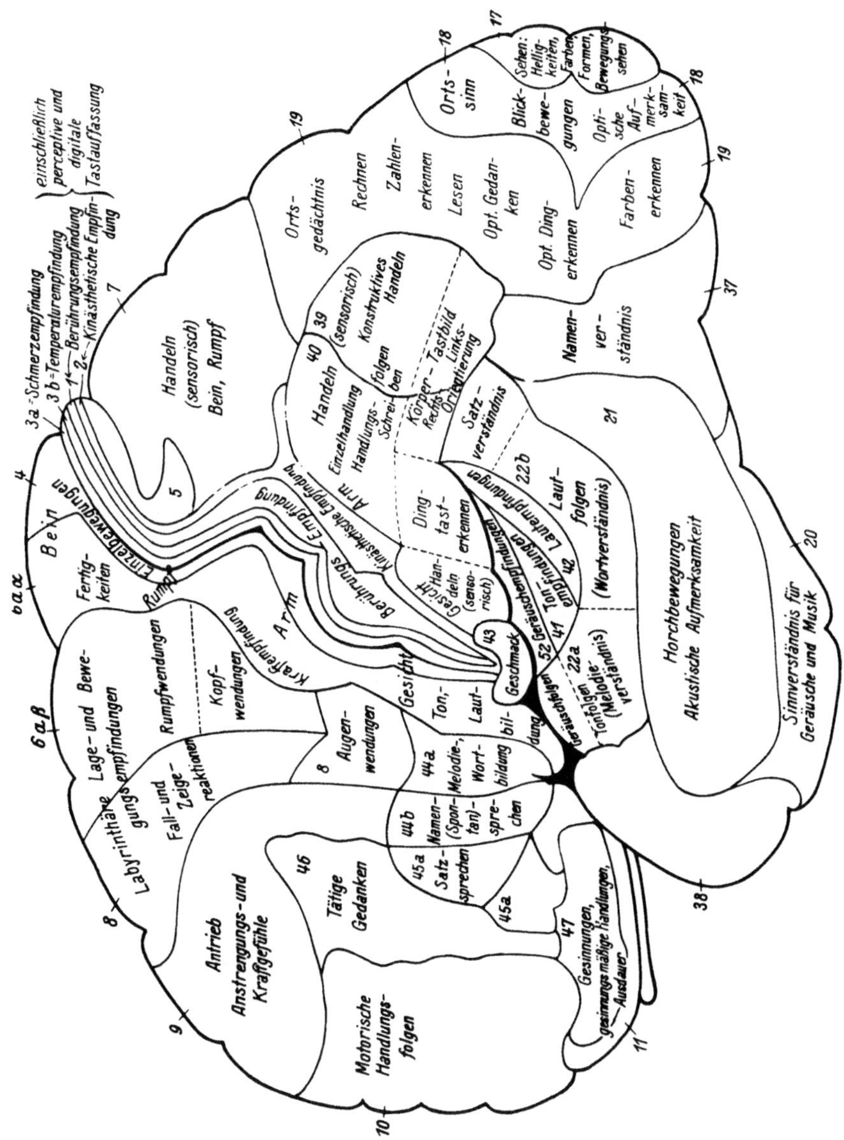

Fig. 3. Cortical localization of function according to Kleist [32].

'faculties' on the surface of the brain, as well as on the skull. As will be seen, each mental ability and personality trait finds a precise locus in a discrete area of the brain.

With the exception of his localization of speech in the frontal lobes, Gall's allocation of specific traits to specific parts of the brain was not taken seriously in scientific circles. However, his fundamental thesis that the human brain is not a unitary equipotential organ was taken very seriously indeed, and it attracted both loyal champions and vigorous opponents. Bitter controversy between 'localizationists' and 'antilocalizationists' over the issue raged for the first two-thirds of the nineteenth century. A series of empirical discoveries by physiologists and physicians during the latter part of the century succeeded in relating specific behavioral impairments to focal brain lesions in specific parts of the brain, and this more or less settled the issue in favor of the localizationists. At least the doctrine that the human brain is a functionally equipotential organ was demonstrated to be clearly untenable.

Scientific localization of cerebral function, i.e., localization based on controlled clinical and experimental observation, began with Broca's demonstration of an association between aphasic disorder and left hemisphere disease in the 1860's and the discovery by Fritsch and Hitzig of the excitable motor cortex in 1870. These pioneer studies were followed by a substantial volume of investigative work that succeeded in establishing associations between a variety of sensory, perceptual, motor, and mental processes and the activity of different parts of the human brain. The enterprise continues until the present day. Figure 3 shows a modern neurological conception of cerebral localization of function, derived from clinical observation of patients with traumatic brain injury. It is that of Karl Kleist [32], a distinguished clinical neurologist who made a number of important contributions to the field. We note that there is some degree of communality between his schema and that of the phrenologist, Thoré, at least as far as the abilities to be localized are concerned. Both find a place for calculation and the sense of location, for example, although the loci selected are not the same.

These three localizational models – ventricular, phrenological and modern – spanning a time period of over 400 years, obviously differ along the dimension of complexity. While the early schema of Reisch localized a half-dozen functions, Thoré's provides for no less than 35 abilities and traits and Kleist's for an even greater number. They also differ with respect to their empirical underpinnings. While the early ventricular and phrenological models were largely the product of speculation, that of Kleist is based on painstaking clinical examination of a large number of patients. However, all proceed from the same basic assumption, namely, that a specific part of the brain, whether it be a ventricle or a cortical area, is the seat or structural substrate of a specific ability or propensity.

As functional localization became a major topic of neurological inquiry, a few thoughtful neurologists and physiologists perceived the difficulties inherent in this dualistic assumption that equated 'mind' with brain. Writing with specific reference to aphasia, Hughlings Jackson (1835-1911) warned that "to locate the lesion which destroys speech and to locate the center of language is not the same thing." His disciple, Henry Head, dealt with the issue at greater length, pointing out that:

> An act of speech is a march of events, where one changing condition passes insensibly into another. When speech is defective, this easy motion or transition is impeded; one state cannot flow into another because of some mechanical imperfection in the process. The power of finding words, the rhythmic modulation and balance of a phrase, the appreciation of meaning and intention, verbal or general, are thrown into disorder.
>
> The processes which underlie an act of speech run through the nervous system like a prairie fire from bush to bush; remove all inflammable material at any one point and the fire stops. So, when a break occurs in the functional chain, orderly speech becomes impossible, because the basic physiological processes which subserve it have been disturbed. The site of such a breach of continuity is not a "centre for speech," but solely a place where it can be interrupted or changed ([26], p. 474).

Similarly, the physiologist, Jacques Loeb (1859-1924), emphasized that the sophisticated efforts of his scientific colleagues in the 1870's and 1880's to localize specific mental functions in discrete cortical areas were, in principle, not different from the earlier efforts of theologians to identify the seat of the soul. He pointed out that the concept of 'function' is only a short-hand term for a whole range of events, and that "cerebral localization" consists in describing how the course of events is changed by induced changes in the nervous system.

> The main task of the physiology of the nervous system lies, it seems to me, in the qualitative and quantitative analysis of the disturbances which follow injuries of the central nervous system. The means and methods of physiology are adequate to such a task and such a phrasing of the question. However, when the question is in the form of: Where does the soul reside? Where does intelligence reside?, it is metaphysical. In contrast, the scientific form of the question is: What differences in quality and quantity of disturbance occur when one removes various parts of the central nervous system under approximately the same conditions? ([35], p. 340).

Thus, in essence, these critics pointed out that the nervous system *is* a system and that the neural loci in it that had been identified as 'centers' subserving behavioral functions are simply crucial links in a complex chain and not the seat of the functions. This is a conception with which all students would agree, but it is one that apparently needs to be kept constantly in mind because, as will be seen, there is a tendency even now to slip back into the simplistic notion that a specific part of the brain is distinctively responsible for a particular function or characteristic, e.g., Wernicke's area in the left temporal lobe for the understanding of language, the frontal lobes for 'awareness of self,' etc. In part, this is due to the clinical diagnostic uses to which our knowledge of cerebral localization is put. Aphasia of the Wernicke type does suggest a lesion in the posterior temporal area of the left hemisphere, and certain types of personality change do suggest disease of the frontal lobes. Thus attention is concentrated on these areas, and it is natural to think of them as being the seat of the disturbed functions and not simply links in a complicated chain of events.

II. CONTRALATERAL INNERVATION

Another form of cerebral localization, dealing with the functions of the two hemispheres, is almost as old as that which located mental functions along the longitudinal axis of the brain. Hippocrates observed that a wound on one side of the head causes motor disturbances on the opposite side of the body. Aretaeus of Cappadocia, a physician and medical writer who flourished in the 2nd century A.D., attempted to explain this observation by postulating a crossing of the nerve tracts descending from the brain to the spinal cord. In due time, over the course of many centuries, his conception was validated by both experimental and clinical study and the fact of contralateral innervation was established. This rule that each hemisphere of the brain controls the motor and sensory functions of the opposite side of the body is of interest to us because it carried the implication that the two hemispheres were equipotential. Thus it probably operated to inhibit recognition of the possibility that the two hemispheres might not be equipotential with respect to other functions, particularly a bilaterally innervated function such as speech.

III. THE 'DOMINANT' HEMISPHERE

As everyone knows, hemispheric asymmetry with respect to speech functions was first demonstrated by the French surgeon and anthropologist, Paul Broca (1824-1880). In 1861, he reported the autopsy findings on two patients who had suffered from motor aphasia during life; in both instances, the responsible lesion appeared to be situated in the left frontal lobe [9, 10, 11]. Broca interpreted his findings as supporting Gall's thesis that the seat of language was in the frontal lobes and made no reference to the circumstance that the lesions were left-sided in both instances. However, as he collected additional cases, his attention was drawn to the unilateral nature of the lesions causing speech impairment. Reporting in 1863 on the autopsy findings in eight aphasic patients, he noted that all had lesions in the left frontal lobe and he cautiously added, "I do not dare to draw a conclusion and I await new findings" [12]. One sees here how hesitant Broca was to embrace the revolutionary idea that one cerebral hemisphere could mediate a complex behavioral function while the other did not possess this capacity. However, the 'new findings' were soon forthcoming and in 1865 he enunciated his famous dictum that "we speak with the left hemisphere" [13].

The validity of Broca's generalization was quickly confirmed and the doctrine of hemispheric cerebral dominance was born. Shortly afterwards, a number of clinical observers (including Broca himself) added an important qualification: left hemisphere representation for speech held only for right-handed persons; in left-handers the right hemisphere appeared to be dominant for language functions. These discoveries led to a revolution in physiological and medical thinking. From a physiologic standpoint, the reality of cerebral localization of function was established. From a medical standpoint, aphasia was transformed from a minor curiosity to an important symptom of focal brain disease.

During the period between Broca's discovery and the end of the 19th century, the concept of left hemisphere dominance (in right-handed persons) was applied only to language functions. However, in the early decades of the 20th century the concept was significantly extended to cover other aspects of behavior and cognition. In 1900, the Berlin neurologist, Hugo Leipmann [34] defined 'apraxia' (i.e., inability to perform a purposeful motor act on command or by imitation) as a distinctive category of behavioral deficit that might be shown by patients with brain disease. He related the deficit to disease of the left hemisphere, a correlation that was fully confirmed by later study. Aphasia specialists such as Pierre Marie [36], Henry Head [26], and Kurt Goldstein [23, 24] strongly emphasized the importance of intellectual impairment as an inherent component of aphasic disorder. Head explicitly defined aphasia as a primary disorder of "symbolic formulation and expression" and not "isolated affections of speaking, reading and writing." Goldstein conceived of at least certain forms of aphasia as being essentially an impairment in abstract thinking that produces altered speech just as a schizophrenic thinking disorder may produce altered speech. This point of view which, to be sure, had already been taken by earlier clinicians, e.g., Trousseau [49] and Jackson [29], fostered the conclusion that the left hemisphere was 'dominant' for higher-level intellectual functions as well as for language in the strict sense.

A third development of a rather different character also significantly extended the concept of left hemisphere 'dominance'. In the 1920's the Viennese neurologist, Josef Gerstmann [21], described 'finger agnosia,' i.e., inability to identify either one's own fingers or those of the examiner. Combining this rather peculiar deficit with three other types of behavioral impairment (namely, defects in discrimination between right and left, in calculation and in writing) into a syndrome, he maintained that this syndrome occurred as a consequence of disease of the left hemisphere. His explanation for the phenomenon, which was widely accepted, was that certain aspects of the 'body-image' found their cerebral representation in the parieto-occipital area of the left hemisphere.

The effect of these observations and inferences was to establish the left hemisphere as the dominant or major hemisphere in the mediation of not only language but also conceptual thinking, certain types of skilled motor activity and orientation to one's body. Thus, as of 1940, cerebral dominance was a unilateral concept for most students of the nervous system. It meant dominance of the left hemisphere.

IV. THE 'MINOR' HEMISPHERE

The notion that the left hemisphere was the dominant or major one implied that its counterpart was the subordinate or minor hemisphere. These terms, 'subordinate' and 'minor', meant that, while the two hemispheres had certain functions in common, such as the contralateral control of sensation and movement, the right hemisphere had no unique functional properties such as had been demonstrated for the left hemisphere.

However, all through this period, a few clinicians, the first of whom was Hughlings Jackson [28, 29, 30], rejected this one-sided formulation of the nature of hemispheric cerebral dominance and insisted that the right hemisphere also served distinctive purposes in the mediation of mental activity (vide [3]). They believed that the behavior of their patients with disease of the right hemisphere showed evidence of a fundamental disorder in visual perception and spatial thinking as reflected in their difficulties in identifying persons, locating objects in space, finding one's way from one place to another, and dressing oneself. Jackson called this assemblage of disabilities "imperception". Others labeled it "visual disorientation". An American ophthalmologist postulated a "geographic centre" in the right side of the brain for the purpose of recording "optical images of locality, analogous to the region of Broca for that of speech on the left side in right-handed persons" ([19], p. 54).

These early observations had virtually no impact on thinking in the field. It was only when large-scale study directly after World War II showed unequivocally that a number of behavioral deficits was encountered with considerably higher frequency in patients with lesions of the right hemisphere as compared to those with left hemisphere disease that conceptions about the nature of hemispheric cerebral dominance were modified. This modification consisted essentially in the addition of a new dimension to the concept of 'dominance'. It was no longer concerned with the left hemisphere alone but rather with the distinctive functions of each hemisphere. As a consequence, the term 'dominance' has seemed to be a less appropriate designation for the actual state of affairs than are the more neutral designations of 'asymmetry of hemispheric function' or 'complementary specialization'.

Developments in the field during the past 20 years have taken a number of directions and forms. There has been intensive study of patients with brain disease in an effort to define the scope and the nature of the cognitive and emotional changes associated with lesions of the right hemisphere. The linguistic and intellectual alterations associated with disease of the left hemisphere have come under close scrutiny. The relationship of handedness to hemispheric asymmetry in cognitive function, an intriguing and puzzling problem, has been examined. An extremely important methodological advance has been achieved by the application of experimental techniques to the study of the hemispheric contribution to perception in normal subjects. This has made it possible to validate the inferences derived from study of patients with brain lesions by complementary observations on persons with intact brains, and conversely to determine whether experimental results with normal subjects find a counterpart in analogous investigation of patients. Finally, there have been attempts to understand the nature of the interrelations between the two hemispheres since it is obvious that, however different the two hemispheres may be, they function together. I shall try to present a picture of our current understanding of what has proved to be a rather complicated issue by sketching some of these developments.

V. THE FUNCTIONAL PROPERTIES OF THE RIGHT HEMISPHERE

Clinical study has shown that a relatively large number of behavioral deficits of a seemingly diverse character and involving the three major sensory modalities – vision, audition and touch – occur with significantly greater frequency in patients with lesions of the right hemisphere than in those with lesions of the left hemisphere. In the visual modality one finds disturbances in the localization of objects in space [18], in constructional activity such as block building and drawing [2], in fine visual discrimination [37], in stereoscopic vision (14), and in the recognition of faces [5]. In the auditory modality, one finds impairment in the discrimination of pitch, tone quality, and temporal patterns of tones [38]. In the tactile modality, one finds impairment in the perception of the direction of linear stimulation applied to the skin surface [15] and in tactile maze learning [17].

The inferential leap from symptom to function, i.e., the conclusion that the intact right hemisphere plays a special role in the mediation of these performances because of the observation that failure in them is associated with disease of that hemisphere, has received substantial support from complementary studies of subjects with intact right hemispheres. For example, it has been found that the recognition of faces is more accurate in the left visual field, information from which projects directly to the right hemisphere, than in the right visual field, information from which projects directly to the left hemisphere [42, 27]. Similarly, tactile perception of direction is more accurate in the left hand, information from which projects directly to the right hemisphere, than in the right hand, which is innervated by the left hemisphere [4].

Nevertheless, it would be fallacious to suppose that the right hemisphere is 'dominant' for any of these performances in the same sense that the left hemisphere is 'dominant' for language activity. The fact is that, although patients with right hemisphere disease may show a remarkably high frequency of impairment on these performances, a significant proportion of patients with left hemisphere disease also show impairment. A comparison between the degree of relationship of aphasia to side of lesion and the degree of relationship of a 'right hemisphere' deficit such as visuoconstructional disability to side of lesion will make this point clear.

Table IA shows the results of two large-scale studies of aphasic patients who were right-handed and whose cerebral lesions were apparently confined to a single hemisphere [16, 43].

The question at issue was how many patients would prove to have lesions of the left hemisphere and how many would be found to have right hemisphere lesions. As will be seen, the proportion of aphasic patients with lesions of the right hemisphere was quite small – 6% in the study of Conrad and less than 2% in the study of Russell and Espir. Thus it is clear that the occurrence of so-called 'crossed aphasia' (i.e., aphasic disorder in a right-handed patient with a right hemisphere lesion) is a very uncommon event. Moreover, in view of the possibility of error in classifying some patients as right-handed or in concluding that the lesion was strictly unilateral in nature, even these small proportions are quite possibly an overestimate of the true frequency of right hemisphere 'dominance' for language in right-handed persons. Given these figures and circumstances, the characterization of the left hemisphere as 'dominant' for language rests on a secure empirical foundation.

In contrast, the situation with respect to a presumed right hemisphere deficit such as visuoconstructional disability is quite different. Table IB shows the findings derived from the data of two studies [1, 2] indicating the side of lesion in 57 patients whose lesions were apparently confined to a single hemisphere and whose performance on a representative visuoconstructional task (three-dimensional block construction) was grossly defective. As will be seen, a substantial majority of the patients proved to have lesions of the right hemisphere and this is consistent with expectations. However, about 30% of the group had lesions of the left hemisphere. Thus, it is clear that there is a significant hemispheric difference in the mediation of this visuoconstructional performance, but it is equally clear that this difference is not of the same order as that which is found for language functions.

TABLE I.

Aphasia and Constructional Apraxia Compared

A. Aphasia

Side of lesion in right-handed aphasic patients
with unilateral brain lesions

	Conrad (1949)	Russell and Espir (1961)
Total number of patients	186	189
Number with *left* hemisphere lesions	175	186
Number with *right* hemisphere lesions	11	3
Percentage with right hemisphere lesions (i.e., 'Crossed Aphasia')	5.9%	1.6%

B. Constructional Apraxia

Side of lesion in right-handed apraxic patients
with unilateral brain lesions

	Arrigoni and De Renzi (1964)	Benton (1967)
Total number of patients	25	32
Number with *right* hemisphere lesions	17	23
Number with *left* hemisphere lesions	8	9
Percentage with left hemisphere lesions (i.e., 'Crossed Constructional Apraxia')	32%	28%

These results are more or less representative of what one finds with the other nonverbal test performances. They pose a formidable problem of interpretation. Do the findings mean that both hemispheres mediate the functions underlying task performance, the right hemisphere being the more important? Is it because the tests that have been contrived can be performed in different ways, i.e., on a perceptual-intuitive basis by the right hemisphere or on a verbal encoding basis by the left hemisphere? Is it possible that the functions are localized differently in the two hemispheres, with a more diffuse representation in the right hemisphere that increases the probability of any given lesion disturbing the function?

These are some of the possibilities that have been advanced and which investigators are currently exploring. In any case, it appears that the 'dominance' of the right hemisphere in respect to certain functions and activities is of a somewhat limited character.

VI. THE RIGHT HEMISPHERE AND LANGUAGE

We have seen that aphasic disorder is almost invariably associated with disease of the left hemisphere in right-handed patients and this indicates the overwhelming importance of the left hemisphere for language functions. However, it does not necessarily imply that the right hemisphere plays no role whatever in mediating language behavior. And, in fact, clinical neurologists have always been inclined to believe that the right hemisphere does participate in language behavior, at least under certain circumstances. There is, first of all, the circumstance that aphasic patients are rarely mute. Their speech may be disordered and they may not be able to communicate with precision, but they are usually capable of uttering automatic, interjectional and emotional speech. That such speech is mediated by the unaffected right hemisphere is indicated by the observations that patients who have had the entire left hemisphere removed engage in it [44], and that pharmacologic inactivation of the right hemisphere in aphasic patients abolishes it [31]. There is also evidence that mechanisms in the right hemisphere in right-handed adults can provide a limited capability for the understanding of oral and written language [44, 20].

Thus, despite its absolutely crucial role in the mediation of speech, the left hemisphere still shares this function with the right hemisphere to a limited degree. Consideration of some aspects of the development of hemispheric cerebral dominance for language in children may shed some light on the reasons for this limited sharing of function. Although most human infants are destined to be left-hemisphere dominant for speech, this focalization of function develops only very gradually during the first 15 years of life. Evidence from several sources indicates that the young child is more or less bicerebral in respect to the exercise of language functions, and that unilateral cerebral dominance for language in older children is less fixed than in adults with the capacity for a shift of the functional representation of language from one hemisphere to another. This plasticity is lost with advancing age, and language functions become irrevocably focalized in the left hemisphere. However, the right hemisphere retains some remnants of its original capacity to mediate linguistic behavior. What is rather surprising is how little it retains.

VII. HANDEDNESS AND CEREBRAL DOMINANCE

The relationship between hand preference and hemispheric specialization of cognitive function presents a fascinating (and puzzling) problem to students of the human nervous system. It will be recalled that directly after Broca discovered the correlation between aphasia and disease of the left hemisphere, he and others noted that left-handed patients did not follow the rule. Some were aphasic as an apparent consequence of right hemisphere disease. Others, who harbored a lesion in the left hemisphere that could have caused aphasia in a right-handed person, showed no signs of language disorder. These observations formed the basis of the doctrine that the right hemisphere is dominant for language in left-handed persons in the same way as the left is in right-handed persons.

This symmetrical model linking handedness to hemispheric specialization for language was accepted by clinical neurologists for almost a century. However, after World War II, systematic studies of left-handed aphasic patients demonstrated that it was not tenable. For these studies found that, in 50-80% of left-handed patients, the aphasic disorder was associated with injury to the left hemisphere. For example, of the 17 left-handed aphasics in Conrad's [16] case material, 10 (59%) proved to have left hemisphere lesions while 7 (41%) had right hemisphere lesions. Similarly, 10 (70%) of the 14 left-handed aphasic patients with unilateral brain wounds studied by Russell and Espir [43] had left hemisphere lesions.

Thus it is quite evident that hemispheric cerebral dominance for language in left-handers is neither the same as, or the mirror-image of, that which is found in right-handers. Hemispheric cerebral dominance for language in a right-handed person is predictable with a fairly high degree of accuracy. It is essentially unpredictable for a left-hander.

There is another important characteristic that distinguished left-handers from right-handers. Both clinical and experimental evidence indicates that many left-handers do not have as strong lateralization of language function and, in fact, that language is bilaterally mediated in some of them [39]. There is suggestive evidence of still another distinctive characteristic. Aphasia in right-handed patients is typically produced by a lesion in a broad, but nevertheless delimited, area of the left hemisphere. In left-handers who have been rendered aphasic by right hemisphere injury, it appears that the lesion is as likely to be outside the conventional speech area as within it [43].

Thus, in some respects, the characteristics of hemispheric cerebral dominance for language in left-handers resembles those of children. In both we find a tendency toward less complete lateralization and greater importance of the right hemisphere than in right-handed adults. There is little to be reported concerning hemispheric specialization with regard to nonverbal abilities in left-handers. The available data are sparse and non-informative.

VIII. HEMISPHERIC SPECIALIZATION AND MODE OF THINKING

A fair amount of recent investigative work and speculation has been concerned not so much with the specific test performances or abilities associated with the activity of each hemisphere as with the question of whether each mediates a characteristic mode of thinking or approach in respect to task performance and problem-solving. The distinction between two basically different modes of thought – a verbal, logical, objective, and analytic mode as opposed to a pictorial, intuitive, subjective, and synthetic mode – is familiar to philosophers and psychologists, and there is much empirical evidence to support the distinction. For example, through the use of factor-analytic techniques, psychologists have been able to demonstrate the existence of a component of intelligence often designated as 'spatial ability,' as contrasted to a 'verbal fluency' component, and they have been able to show that there are differential relations between these components and such factors as sex, occupation, and profile of academic achievement [45].

As we have seen, Hughlings Jackson [28, 29, 30] made this distinction over 100 years ago and suggested that the two hemispheres serve diametrically opposing functions. Some decades later a similar division was made and applied explicitly to brain function by the German neurologist, Rieger (1909), who postulated the existence of two distinct and interacting 'apparatuses of the brain', one mediating verbal-conceptual thinking, the other mediating spatial-practical thinking. He had little to say about the anatomic locus of these mechanisms except to place the spatial-practical apparatus in the hinder part of the brain and the verbal apparatus in the insular region of the left hemisphere.

In our day the idea has been taken up by a number of investigators, most notably by Sperry, Gazzaniga, and Bogen [46, 47, 6, 7, 8, 33] in connection with their outstandingly important studies of patients whose cerebral hemispheres have been disconnected by section of the corpus callosum and the anterior commissure. Their observations have led them to conclude that the two hemispheres mediate quite different cognitive approaches to experienced events. The right hemisphere provides a basis for direct apprehension of the situation as a whole without recourse to detection of specific details or verbal description. In contrast, the left hemisphere provides a basis for the acquisition of knowledge by successive detection of details facilitated by verbal encoding. The first approach has been called 'gestalt-synthetic' and 'appositional'; the second, 'logical-analytic' and 'propositional'. The right hemisphere processes information in parallel, the left hemisphere in series.

Diverse empirical findings can be marshalled in support of the conception. Sperry and Levy [47] note that there are qualitative, as well as quantitative, differences in certain task performances depending upon whether the performance is mediated by the right or the left hemisphere. 'Right hemisphere' performance is quite rapid and nonverbal, pointing to immediate apprehension of the available information. In contrast, 'left hemisphere' performance is more deliberate and accompanied by verbalization. When the left hemisphere is mediating performance, a task that includes features which are easily verbalized is done better than one which does not include these features. On the other hand, the presence or absence of easily verbalized features does not affect performance mediated by the right hemisphere.

This coupling of hemispheric specialization with cognitive typology has stimulated some interesting and far-reaching speculation. No doubt there are individual differences in what Bogen *et al.* [8] have called the *A/P* or appositional-propositional ratio. Some people rely more on the left hemisphere in thinking while others rely more on the right hemisphere. Conceivably there are genetic bases for these differences but almost surely environmental factors also play a determining role. Formal education, with its emphasis on verbal-conceptual modes of thinking, favors the use of the left hemisphere. Life in a slum, or any other milieu that makes demands on quick perception and intuitive judgment, favors the use of the right hemisphere.

Handedness has also been brought into the appositional-propositional dichotomy. As has been pointed out, it appears that language functions are mediated to a significant degree by both hemispheres in some left-handers. This observation has led to the hypothesis that, since the right hemisphere of these individuals has been preempted for the exercise of the verbal-conceptual mode of thought, it cannot serve as adequately to sustain the perceptual-intuitive mode. This hypothesis leads to the prediction that, as a group, left-handers should show a poorer development of perceptual-spatial abilities than of verbal-conceptual abilities. Empirical testing of this prediction to date has produced a mélange of positive and negative results.

Finally, some workers in the field have ventured to draw inferences from the facts of hemispheric asymmetry about the age-old problem of the 'unity of the mind'. Their conclusion has been that the demonstration of two separate and parallel cognitive systems, one in each hemisphere, argues for an essential duality of mind and, hence, that a person's subjective feeling of unity is an illusion [6].

IX. A SUMMING UP

Let us recapitulate the story. It begins with a discovery that was made relatively late in medical history, indeed surprisingly late, considering that the crucial facts were plain enough and yet they escaped the notice of the most astute observers. This discovery of a systematic association between impairment in language function and disease of the left hemisphere led to a revolution in thinking about the human brain and provided the impetus for a determined effort on the part of both clinicians and experimental scientists to understand the nature of its functional organization.

Developments took two parallel courses. One line of study emphasized that the left hemisphere was crucial in the mediation of a number of cognitive functions in addition to language, thus leading to the hierarchic concept of a major dominant hemisphere coupled with a minor subordinate hemisphere. The other line of inquiry adduced evidence, initially quite weak but over the decades steadily gaining cogency, that the right hemisphere was crucial in the mediation of a number of other cognitive activities, thus leading to the more democratic concept of asymmetry of hemispheric function or hemispheric complementarity. That hand preference is intimately connected with hemispheric specialization for language was recognized quite early.

More recent developments have strengthened the concept of asymmetry of hemispheric function. There is now broad acceptance of the generalization that (in right-handed adults) the neural mechanisms underlying non-verbal pictorial thought are predominantly (but probably not exclusively) concentrated in the right hemisphere.

There is much that remains unknown. The nature of the cognitive processes that are altered when the right hemisphere is injured has not been precisely defined. The relationship of hand preference to hemispheric function is not well understood. Very little is known about how the neural mechanisms in the two hemispheres integrate to provide a basis for mentation and behavior. The question of differences in the organization of neural mechanisms within each hemisphere (e.g., more diffuse organization in the right hemisphere than in the left) requires elucidation.

Earlier I mentioned that thoughtful students of the nervous system such as Hughlings Jackson and Jacques Loeb had insisted that the central nervous system must be seen as a system and not as an assemblage of discrete organs, each mediating a discrete mental function. Yet it seems that the latter archaic view is still held by otherwise quite sophisticated investigators. One reads statements such as "the right hemisphere can understand language but it cannot talk," "the right hemisphere as well as the left hemisphere can emote," and "each of us has two minds in one person." No doubt most of these statements can be regarded as harmless metaphors. However, they seem sometimes to be more than that. They seem to reflect a way of interpreting observed facts that emphasizes the functional autonomy of each hemisphere and that diverts attention from the problem of how the activities of the two hemispheres interact and integrate to serve cognitive behavior. This is really the basic question and the one that must be answered if a satisfying understanding of human brain function is to be achieved.

University of Iowa,
Iowa City, Iowa

BIBLIOGRAPHY

1. Arrigoni, G. and De Renzi, E.: 1964, 'Constructional Apraxia and Hemispheric Locus of Lesion', *Cortex* **1**, 170-197.
2. Benton, A.L.: 1967, 'Constructional Apraxia and the Minor Hemisphere', *Conf. Neurol.* **29**, 1-16.
3. Benton, A.L.: 1972, 'The "Minor" Hemisphere', *J. Hist. Med. All. Sci.* **27**, 5-14.
4. Benton, A.L., Levin, H.S., and Varney, N.R.: 1973, 'Tactile Perception of Direction in Normal Subjects: Implications for Hemispheric Cerebral Dominance', *Neurology* **23**, 1248-1250.
5. Benton, A.L. and Van Allen, M.W.: 1968, 'Impairment in Facial Recognition in Patients with Cerebral Disease', *Cortex* **4**, 344-358.
6. Bogen, J.E.: 1969, 'The Other Side of the Brain: II. An Appositional Mind', *Bull. Los Angeles Neurol. Soc.* **34**, 135-162.
7. Bogen, J.E. and Bogen, G.M.: 1969, 'The Other Side of the Brain: III. The Corpus Callosum and Creativity', *Bull. Los Angeles Neurol. Soc.* **34**, 191-220.
8. Bogen, J.E., De Zure, R., Tenhouten, W.D. and Marsh, J.F.: 1972, 'The Other Side of the Brain: IV. The *A/P* Ratio', *Bull. Los Angeles Neurol. Soc.* **37**, 49-61.
9. Broca, P.: 1861, 'Nouvelle Observation d'Aphémie Produite par une Lésion de la Moitié Postérieure des Deuxième et Troisième Circonvolutions Frontales Gauches', *Bull. Soc. Anat. Paris* **6**, 398-407.
10. Broca, P.: 1861, 'Perte de la Parole: Ramolissement Chronique et Destruction Partielle du Lobe Antérieure Gauche du Cerveau', *Bull. Soc. Anthrop. Paris* **2**, 235-238.

11. Broca, P.: 1861, 'Remarques sur le Siège de la Faculté du Langage Articulé, Suivies du'une Observation d'Aphémie', *Bull. Soc. Anat. Paris* **6**, 330-357.
12. Broca, P.: 1863, 'Localisation des Fonctions Cérébrales: Siège du Langage Articulé', *Bull. Soc. Anthropol. Paris* **4**, 200-203.
13. Broca, P.: 1865, 'Sur la Faculté du Langage Articulé', *Bull. Soc. Anthropol. Paris* **6**, 493-494.
14. Carmon, A. and Bechtoldt, H.P.: 1969, 'Dominance of the Right Cerebral Hemisphere for Stereopsis', *Neuropsychologia* **7**, 29-39.
15. Carmon, A. and Benton, A.L.: 1969, 'Tactile Perception of Direction and Number in Patients with Unilateral Cerebral Disease', *Neurology* **19**, 525-532.
16. Conrad, K.: 1949, 'Über äphasische Sprachstörungen bei hirnverletzten Linkshandern', *Nervenarzt* **20**, 148-154.
17. Corkin, S.: 1965, 'Tactually-Guided Maze Learning in Man: Effects of Unilateral Cortical Excisions and Bilateral Hippocampal Lesions', *Neuropsychologia* **3**, 339-352.
18. De Renzi, E. and Faglioni, P.: 1967, 'The Relationship Between Visuo-Spatial Impairment and Constructional Apraxia', *Cortex* **3**, 327-342.
19. Dunn, T.D.: 1895, 'Double Hemiplegia with Double Hemianopsia and Loss of Geographic Centre', *Trans. Coll. Physicians Philadelphia* **17**, 45-56.
20. Gazzaniga, M.S.: 1970, *The Bisected Brain*, Appleton-Century-Crofts, New York.
21. Gerstmann, J.: 1924, 'Fingeragnosie: eine umschriebene Störung der Orientierung am eigenen Körper', *Wien. Klin. Wschr.* **37**, 1010-1012.
22. Gerstmann, J.: 1927, 'Fingeragnosie und isolierte Agraphie – ein neues Syndrom', *Z. Neurol. Psychiat.* **108**, 152-177.
23. Goldstein, K.: 1924, 'Das Wesen der amnestischen Aphasie', *Schweiz. Arch. Neurol. Psychiat.* **15**, 163-175.
24. Goldstein, K.: 1926, 'Über Aphasie', *Schweiz. Arch. Neurol. Psychiat.* **19**,3-38.
25. Guainerio, A.: 1481, *Opera Medica*, Antonius de Carcano, Pavia.
26. Head, H.: 1926, *Aphasia and Kindred Disorders of Speech*, The University Press, Cambridge.
27. Hilliard, R.D.: 1973, 'Hemispheric Laterality Effects on a Facial Recognition Task in Normal Subjects', *Cortex* **9**, 246-258.
28. Jackson, J.H.: 1864, 'Clinical Remarks on Cases of Defects of Expression (by Works, Writing, Signs, etc.) in Diseases of the Nervous System', *Lancet* **2**, 604-606.
29. Jackson, J.H.: 1874, 'On the Nature of the Duality of the Brain', *Med. Press Circ.* **17**, 19, 41, 63 (reprinted in *Brain* **38**, (1915), 80-103).
30. Jackson, J.H.: 1876, 'Case of Large Cerebral Tumour Without Optic Neuritis and With Left Hemiplegia and Imperception', *Roy. Ophthal. Hosp. Rep.* **8**, 434-444.
31. Kinsbourne, M.: 1971, 'The Minor Cerebral Hemisphere as a Source of Aphasic Speech', *Arch. Neurol.* **25**, 302-306.
32. Kleist, K.: 1934, *Gehirnpathologie*, Barth, Leipzig.
33. Levy, J., Trevarthen, C. and Sperry, R.W.: 1972, 'Perception of Bilateral Chimeric Figures Following Hemispheric Deconnexion', *Brain* **95**, 61-78.
34. Leipmann, H.: 1900, 'Das Krankheitsbild der Apraxie (Motorischen Asymbolie) auf Grund eines Falles von einseitiger Apraxie', *Mschr. Psychiat. Neurol.* **8**, 15-44, 102-132, 182-197.

35. Loeb, J.: 1886, 'Beiträge zur Physiologie des Grosshirns', *Pflügers Arch. Ges. Physiol.* **39**, 265-346.
36. Marie, P.: 1906, 'La Troisième Circonvolution Gauche ne Joue Aucum Rôle Spécial dans la Fonction du Langage', *Semaine Médicale* **26**, 241-247. (English translation in M.F. Cole and M. Cole: 1971, *Pierre Marie's Papers on Speech Disorders*, Hafner, New York).
37. Meier, M.J. and French, L.A.: 1965, 'Lateralized Deficits in Complex Visual Discrimination and Bilateral Transfer of Reminiscence Following Unilateral Temporal Lobectomy', *Neuropsychologia* **3**, 261-272.
38. Milner, B.: 1962, 'Laterality Effects in Audition', in V.B. Mountcastle (ed.), *Interhemispheric Relations and Cerebral Dominance*, Johns Hopkins Press, Baltimore.
39. Milner, B.: 1973, 'Hemispheric Specialization: Scope and Limits', in F.O. Schmidt and F.G. Worden (eds.), *The Neurosciences: Third Study Program*, M.I.T. Press, Boston.
40. Pagel, W.: 1958, 'Medieval and Renaissance Contributions to Knowledge of the Brain and its Functions', in F.N.L. Pointer (ed.), *The History and Philosophy of Knowledge of the Brain and its Functions*, Blackwell, Oxford.
41. Reisch, G.: 1504, *Margarita Philosophica*, 2nd ed., Freiburg.
42. Rizzolati, G., C. Umiltà, and G. Berlucchi: 1971, 'Opposite Superiorities of the Right and Left Cerebral Hemispheres in Discriminative Reaction Time to Physiognomic and Alphabetical Material', *Brain* **94**, 431-442.
43. Russell, W.R. and Espir, M.L.E.: 1961, *Traumatic Aphasia*, Oxford University Press, London.
44. Smith, A.: 1966, 'Speech and Other Functions After Left (Dominant) Hemispherectomy', *J. Neurol. Neurosurg. Psychiat.* **29**, 467-471.
45. Smith, I.M.: 1964, *Spatial Ability: Its Educational and Social Significance*, Robert R. Knapp, San Diego.
46. Sperry, R.W., Gazzaniga, M.S. and Bogen, E.: 1969, 'Interhemispheric Relationships; the Neocortical Commissures; Syndromes of Hemisphere Disconnection', in P.J. Vinken and G.W. Bruyn (eds.), *Handbook of Clinical Neurology*, Vol. 4, North-Holland Publishing Company, Amsterdam.
47. Sperry, R.W. and Levy, J.: 1970, 'Mental Capacities of the Disconnected Minor Hemisphere Following Commisurotomy', presented in symposium on "Asymmetrical Function of the Human Brain," Annual Convention, American Psychological Association, Miami, Florida.
48. Thoré, T.: 1836, *Dictionnaire du Phrénologie, et de Physiognomie*, à la Librairie Usuelle, Paris.
49. Trousseau, A.: 1864, 'De l'Aphasie, Maladie Décrite Récemment sous le Nom Impropre d'Aphémie', *Gaz. Hôp. Paris* **37**, 13-14, 25-26, 37-39, 48-50.

H. TRISTRAM ENGELHARDT, JR.

REFLECTIONS ON OUR CONDITION: THE GEOGRAPHY OF EMBODIMENT

The papers by William Bynum and Arthur Benton provide sketches of the history of reflections by physicians and neural scientists about the relationship of mind and body. For many reasons we puzzle concerning what our minds have to do with our bodies. We know ourselves as fabrics of volitions, desirings, and cognitions that have a history and a place. I am in the here and now of my body, one of the physical objects of this world. In particular, the brain has shown itself to be my location, *par excellence*. Thus, my first-person acquaintance with my feelings, longings, and willings is confronted with a radical dependence on the central nervous system, which, in contrast, can be studied as can any other object in the world. Because of the contrast of the subjective and the objective, and because of their intimate coincidence in the brain, one has that recurring set of puzzles which is termed the mind-body problem. The experienced sense of mind and that of body contrast, yet they are intimately interrelated: the very sense of the mental in this world presupposes a body [4].

The history of the neural sciences is in part the history of an attempt to resolve a set of problems concerning the way in which the relation of mind and body should be viewed. One, for example, finds modern analogues of scholastic disputes concerning the localization of the soul in the body in contemporary controversies about the brain-oriented concept of death: the puzzle is the point at which the brain ceases to sustain the mind. Such reflections presuppose that brain structure of a certain sort is requisite for there being a mind of a certain sort. The history of the neural sciences has in part focused on clarifying that presupposition, both in terms of factual issues (i.e., what parts of the brain are necessary for what human activities), and in terms of theoretical issues (i.e., how should one talk about the functions of the brain so that a coherent account of the facts would be possible).

S. F. Spicker and H. T. Engelhardt, Jr. (eds.), Philosophical Dimensions of the Neuro-Medical Sciences, 59–68.

William Bynum has given a very useful account in this mode, contending that (1) natural theological faith in the concordance of structure and function, (2) the development of a tradition of sensory-motor physiology, and (3) the common ascription of minds to animals by Franz Josef Gall and Pierre Flourens determined central elements of the development of neurophysiology in the early nineteenth century. Moreover, William Bynum gave a useful examination of three cardinal propositions of Gall's phrenology: (1) that the brain is the organ of the mind, (2) that the brain is an aggregate of several organs, each subserving a distinct mental faculty, and (3) that the size of the brain is, all else being equal, an index of mental function.

In my brief remarks I will focus on the second of each of these sets of three propositions: the significance of the tradition of sensory-motor physiology and Gall's thesis that the brain is an aggregate of organs. It is in terms of these that I believe that an answer can be given to William Bynum's unanswered question of why Flourens dedicated to Descartes his book criticizing phrenology, *Examen de la phrénologie.* [5] In considering these issues, I will be brought to say a number of things about Hughlings Jackson, who first addressed many of the issues raised by Professor Benton concerning the concept of hemispheric dominance.

Of all the important conceptual developments of nineteenth century neurophysiology, the development of a sensory-motor idiom was one of the most salient. It signaled a departure from the use of faculty psychology in describing neural function. It also involved a departure from seeing neurophysiology as the study of the function of the physical instruments of the mind, to seeing neurophysiology as the study of sensory-motor integration. I do not mean to imply that there were not materialistic accounts of mind and of the nervous system before the nineteenth century. What was new was a fairly complete idiom free of psychological terms, which terms had been translated into the language of sensory-motor integration. Gall and Flourens played quite different roles in this development.

On the one hand, as Bynum indicates, Gall remained within the idiom of mental faculties even though he multiplied them considerably. Gall continued to conceive of the function of the nervous system in mental terms, proposing organs for faculties such as "love of offspring." This mentalistic element of Gall's interests is somewhat obvious. What is not obvious, and a point that Bynum fails to stress sufficiently, is that Gall, almost paradoxically, was interested in seeing this psychological idiom in physiological terms.

Gall, for example, arrayed himself against those who held that "the faculties of the mind are by no means an object of physiology," ([7], p. 73) and against those such as Flourens who for "metaphysical" reasons wanted to argue for the unity of neurophysiological functions on the basis of the unity of mind. In contrast, Gall wished to give an account of the function of the brain in terms of its structure so that one could speak of "the received impression [being] transformed into perception, recollection, image, judgment, in the same manner that we can see physiologically the alimentary paste successively transformed into chyle, blood, [etc.]." ([7], p. 71). Moreover, in his approach Gall made claims to being fully empirical. "I have never permitted myself to be influenced by reasoning *a priori*... . I preferred to abandon myself entirely to observations." ([7], p. 129). Gall conceived of himself as an empiricist arrayed against rationalists such as Flourens.

On the other hand, Flourens wanted to preserve the integrity of the psychological idiom and he wished to do so because of philosophical, in particular moral, considerations. Flourens saw Gall as undermining the concepts of free will, ([6], pp. 42-43) of the unity of the mind, ([6], pp. 44-45, 52-53) of God, ([6], p. 55) and of the immortality of the soul ([6], p. 123). To secure his position Flourens argued in a Cartesian fashion from the experienced unity of mind and its conceptual distinctness to its real distinctness from the body. For example, he contended that "Consciousness which gives me the *unity* of the *moi*, gives me not less assuredly the *continuity* of the *moi*." ([6], p. 123). On this point he quoted Descartes: "We do not conceive of any body except as divisible; whereas the human mind cannot conceive of itself except as indivisible," ([6], p. 57; [3], p. 51) and argued for the distinctness and unity of consciousness. In short, Flourens could be considered in part a rationalist who wished to secure certain premises about the nature of man through a clear distinction between mind and body, and through a Cartesian doctrine of the interaction of mind and body. There should be no question of why he dedicated his critique of phrenology to Descartes.

Thus, the positions of Gall and Flourens contrast, for with respect to their view of man, Gall was an empiricist and Flourens a rationalist. This may seem to be an unlikely description considering Gall's excesses with his faculty psychology and organology, but his were the sins of a flamboyant and somewhat inconsistent empiricist. It was Flourens who preserved a central place for Cartesian language concerning a soul interacting with a body, while Gall more often saw the brain as itself performing 'mental' functions and this tended to invite a physiological, empirical description of 'mental' functions. Thus, in contrast, Flourens spoke of "the cerebral hemispheres concur[ring], by their whole mass in the full and entire exercise of the intelligence." ([6], p. 34). He spoke of the hemispheres as the seat of the soul and the soul as alone being sentient: "... if it be the *soul* only that feels, *a fortiori*, it is the soul only that *remembers*, [etc.]..." ([6], pp. 26-27). That is not to say that Gall did not continue to talk of the soul as *using* the body.[1] But Gall's general outlook was one of viewing the brain as a mechanism which could be accounted for in physiological terms without recourse to any metaphysical or philosophical premises concerning the nature of consciousness. Flourens, who wished to secure a physiological basis for maintaining that there is a free will, continued to hold that mentalistic and psychological language were distinct and that, in addition, certain neurological functions, such as those performed by the cerebrum, could only be accounted for in mentalistic terms.

In short, William Bynum contributes a very useful account of the often overlooked similarities of Gall and Flourens while slighting a central dissimilarity. He does not address the Cartesian roots of the conflict between Flourens and Gall as the title of his paper implies he would: Flourens was a Cartesian and Gall was not. Gall contributed a non-Cartesian intention to see the physiology of the human brain in biological terms[2] without introducing any a priori psychological unities. In the hands of John Hughlings Jackson, Gall's view of physiology developed into a somewhat Leibnizian account of mind and body within which talk of mind-body interaction ceased, and for which talk of a concomitance (parallelism) of mind and body was substituted, providing a thoroughly physiological idiom for the consideration of brain functions. In short, a consistent development of a sensory-motor physiology required the abandonment of a priori grounds against the organology of Gall and a turn away from Cartesian presuppositions, at least as Flourens saw them. In the history of the neural sciences, perceptive arguments for conceptual consistency in these matters were made by John Hughlings Jackson.

Consideration of Hughlings Jackson's general, philosophical suggestions is omitted by Bynum and Benton, though Professor Benton treats of Jackson on other issues. Hughlings Jackson was a watershed figure. He developed a language of sensory-motor integration for use in describing brain function, including the function of the cerebral hemispheres, a point to which I will return. Jackson wanted to free neurology from the use of mentalistic terms so that truly neurological explanations (i.e., consistently in physical, not mental terms) could be developed, and so that one would not think that neurophysiological functions had been explained by making reference to mental events. He thus warned against use of mentalistic terms in framing physiological explanations, even such as are implied in terms like "psycho-motor," ([20], p. 44) or in statements that a person cannot move because he is unconscious. He termed the latter a "metaphysical explanation." ([9], p. 662; [10], p. 651; [11], p. 447; [19], p. 59). Apropos the issue of hemispheric dominance, he argued against referring to loss of speech as if this concerned the loss of a mental function, when studying the physiology of speech.

Loss of speech is therefore the loss of power to propositionise.... We do not mean, by using the popular term power, that the speechless man has lost any 'faculty' of speech or propositionishing... . There is no 'faculty' or 'power' of speech apart from words revived or revivable in propositions, any more than there is a 'faculty' of co-ordination of movements apart from movements represented in particular ways.... It is well to insist again that speech and words are psychical terms; words have of course anatomical substrata or bases as all other psychical states have. ([13], pp. 312-13; [19], p. 160)

Jackson thus offered a methodological materialism through which he treated the nervous system as a "sensori-motor mechanism" or a "sensori-motor machine [in which] there is nothing going on in it, other than nervous discharges." ([17], p. 37; [19], p. 84). He adopted this view because he held that it was impossible to study the nervous system if one confused psychical and nervous states. ([16], [19], p. 9). This view of the relation of mind and body Jackson termed his doctrine of concomitance; he held, "... first, that states of consciousness (or, synonymously, states of mind) are utterly different fron nervous states; second, that the two things occur together – that for every mental state there is a correlative nervous state; third, that although the two things occur in parallelism, there is no interference of one with the other." ([9], p. 706; [10], p. 742; [11], p. 486; [19], p. 72). In short, Jackson offered a way of sorting out the languages and concerns of neuro-physiology and of psychology so that each could be treated in its own right. Jackson recognized mind and body as two different categories of reality, not as two *things* that interacted [4]. Importantly, Jackson's exorcism of mentalistic terms from the vocabulary of neurology did not imply that he lacked interest in psychological questions. In fact, he engaged in significant psychological speculation.

These reflections concerning Hughlings Jackson are important to an understanding of Professor Benton's suggestions concerning hemispheric dominance, especially when he begins to talk of what "mental activity" the hemispheres might "mediate." ([1], p. 45). Benton is not at all clear about the significance of the concept of dominance. He suggests that the hemispheres are concerned with different modes of thinking, ([1], pp. 51-53) and then remarks that Jackson suggested that the hemispheres serve different functions. Jackson, though, became very careful in his use of concepts such as 'function,' to which he denied any mentalistic significance, holding that it was a "physiological term; it has to do with the dynamics of the nervous system, with things physical only." ([17], [19], p. 84). In particular, Jackson was wary of phrases implying that the hemispheres were "mediating verbal-conceptual thinking." ([1], p. 52). Which is to say, Professor Benton does not distinguish the issue of the dominance of one mode of thinking over another from the dominance of one mode of sensory-motor integration over another. But this distinction is crucial. It is one thing to decide on the unity of the sensory-motor integration which occurs in the cerebrum and whether one hemisphere is in some sense preeminent in general, or preeminent with respect to certain modes of sensory-motor integration. It is another to decide whether persons with isolation of the function of the two hemispheres experience a disunity of consciousness, or even whether any*one* could experience a disunity of consciousness. That is, one might instead decide there was someone else in his other hemisphere. ([8], p. 107). In any event, the meaning of 'dominance' in each case is quite different: the first question concerns sensory-motor integration, the second concerns the nature of mind. The history of the discussion of the unity of the hemispheres has, in short, suffered from its participants' not having consistently followed Hughlings Jackson's advice to distinguish between physiological and psychological inquiries.

On this point, Professor Benton remarks that some investigators have used information concerning hemispheric dominance to make conclusions about the unity of mind and have, indeed, come to assert that there are two minds in one person ([1], p. 53) – a phrase, I might add, which sounds as mysterious as any doctrine of the Trinity. Benton gives reference in this regard to Joseph Bogen, who has argued that there are in each person a propositional and an appositional mind ([2], p. 156). If there is to be a solution to such puzzles, one must attend carefully to the meaning of the inquiry: is one interested in deciding how cerebral hemispheres integrate their functions, or how patients who undergo corpus callosum sections experience their situations? If one is interested in the second inquiry, one must decide on the sense of such peculiar notions as having two minds who would have them?), but that takes us beyond the scope of my remarks here.

This problem of the unity of the person, with which Professor Benton ends his paper, brings us back to the question of the relationship between the unity of consciousness and the unity of the cerebrum, the question treated by Flourens. Hughlings Jackson gave a helpful answer to the question in his résumé of his original discussions of these issues, which he titled suggestively, "On the Nature of the Duality of Brain" ([14], pp. 19, 41, 63; [15], pp. 80-103; [19], pp. 129-145). In introducing distinctions between the voluntary speech of the dominant hemisphere and the automatic speech of the non-dominant hemisphere, he apologized for the fact that "the words voluntary and automatic [have psychological implications], but they are also used physiologically" ([14], p. 22 n. 2; [15], p. 84 n. 2; [19] p. 132, n. 2). Later, he used his distinction of voluntary and involuntary speech to speak of the "Duality of Mental Processes" ([12], p. 526; [19], p. 227). Jackson's solution seems to be that the integration of the hemispheres is like the integration of all sensory-motor function: it is a coordination of sensory-motor integration ([12], pp. 208-209, 358-59; [19], pp. 220-26). Mental integration, or the coordination of mental functions, will exist (when it does exist) *pari passu* with such sensory-motor integration. If certain lesions should cause a separation of the functions of the hemispheres, it would simply be another example of the many pathological disorganizations that occur and that release otherwise integrated functions of the brain. The question of whether consciousness then becomes double would be a phenomenological issue, a question of describing what the split brain state is like for those persons who experience it ([18], [19], p. 96). A study of the history of the concept of hemispheric dominance might contribute by pointing out how little these psychological and physiological issues have been distinguished, and as a result contribute both to a clearer physiology and to a phenomenology of the split brain state.

In conclusion, the mind-body problem continues to be for us, as it was for Descartes, an issue of interest because of its philosophical, as well as its empirical, implications. The history of neurophysiology and neurology provided by Bynum and Benton indicates the role of conceptual frameworks, paradigms if you will, in the neural sciences. Gall's phrenology stimulated interest in a localizing geography for the embodiment of mind, Flourens and Gall invited the study of the continuity of neural structure in animals and humans, and Hughlings Jackson provided a developed idiom of sensory-motor integration supporting the pursuit of neurological questions without confusing them with psychological issues.

Finally, how one answers the basic question of the relationship of mind and brain has implications which bear immediately on the practice of medicine, for the answers determine how one views the significance of neurological conditions for those who have them, patients. To ask what corpus callosum sections do to the unity of consciousness is to ask what effects they have upon the patients subjected to them. To decide whether a hemisphere can be removed in order to remove a tumor is to decide whether that hemisphere is a necessary condition for the person involved continuing as a person. That is, to use Jacksonian language, one is forced to inquire what parts of the brain are the physical substrates of consciousness, and what physiological integration is the substrate of the integration of consciousness. Because it is the implications of neurosurgery for the conscious life of the individual undergoing the surgery that determine the success of the surgery (e.g., if the individual lives but is no longer conscious, in what sense does he remain a person to benefit from the surgery?), it becomes necessary to determine what alterations of physical structure alter the life of mind and what alterations are made. These issues concern the geography of embodiment, the ways in which the mind and the particular abilities of the mind require the functions of particular areas of the brain. The history of the neural sciences as presented indicates that Gall was in large measure correct: we are, as minds, laid out upon the shifting geography of brain function. Even if the lines are often imprecise and moving, like the boundaries between contested territories, we as minds have our here and now in this world through the here and now of our bodies – which is really to lead us from these historical issues to the issues raised in the papers below.

University of Texas Medical Branch,
Galveston, Texas

NOTES

[1] "... I have proved that... in this life, the manifestation of none of these dispositions is possible, without the aid of material instruments" ([7], p. 1).

[2] R.M. Young provides a very imaginative treatment of the development of a sensory-motor idiom in describing neurophysiological function [21].

[3] It is worth noting that Jackson was not impressed by the unity of consciousness in any Cartesian sense. He held that '*I*' stood for an activity unknowable in itself, 'subject consciousness', which is known only in its deliverance, 'object consciousness' ([18], [19], p. 96).

BIBLIOGRAPHY

1. Benton, A.: 'Historical Development of the Concept of Hemispheric Cerebral Dominance', in this volume, pp. 35-57.
2. Bogen, J.E.: 1969, 'The Other Side of the Brain, II: An Appositional Mind', *Bulletin of the Los Angeles Neurological Societies* **34**, 135-162.
3. Descartes, R.: 1658, 'Meditationes De Prima Philosophia', Meditation VI, in *Opera Philosophica*, Editio Ultima, Johannem Janssonium Juniorem, Amsterdam, 1656-1658.
4. Engelhardt, H.T., Jr.: 1973, *Mind-Body: A Categorial Relation*, Martinus Nijhoff, The Hague.
5. Flourens, M.J.P.: 1845, *Examen de la phrénologie*, 2nd ed., Paulin, Paris.
6. Flourens, M.J.P.: 1846, *Phrenology Examined*, (transl. by C.L. Meigs), Hogan and Thompson, Philadelphia.
7. Gall, F.J.: 1835, *On the Functions of the Brain and of Each of Its Parts*, Vol. 3, Marsh, Capen and Lyon, Boston.
8. Gazzaniga, M.S.: 1970, *The Bisected Brain*, Appleton-Century-Crofts, New York.
9. Jackson, J.H.: 1884, 'Evolution and Dissolution of the Nervous System', *British Medical Journal* **1**, 591-93, 660-63, 703-707.
10. Jackson, J.H.: 1884, 'Evolution and Dissolution of the Nervous System', *Lancet* **1**, 555-58, 649-52, 739-44.
11. Jackson, J.H.: 1884, 'Evolution and Dissolution of the Nervous System', *Medical Times and Gazette* **1**, 411-13, 445-48, 485-87.
12. Jackson, J.H.: 1868, 'Notes on the Physiology and Pathology of the Nervous System', *Medical Times and Gazette* **2**, 177-79, 208-209, 358-59, 526-28, 696.
13. Jackson, J.H.: 1878-1879, 'On Affections of Speech from Disease of the Brain', *Brain* **1**, 304-330.
14. Jackson, J.H.: 1874, 'On the Nature of the Duality of the Brain', *Medical Press and Circular* **1**, 19-21, 41-44, 63-65.
15. Jackson, J.H.: 1915, 'On the Nature of the Duality of the Brain', *Brain* **38**, 80-103.
16. Jackson, J.H.: 1881, 'Remarks on Dissolution of the Nervous System as Exemplified by Certain Post-Epileptic Conditions', *Medical Press and Circular* **1**, 329-332.

17. Jackson, J.H.: 1887, 'Remarks on Evolution and Dissolution of the Nervous System', *Journal of Mental Science* **33**, 25-48.
18. Jackson, J.H.: 1887, 'Remarks on Evolution and Dissolution of the Nervous System', *Medical Press and Circular* **2**, 461-462, 511-513, 586-588, 617-620.
19. Jackson, J.H.: 1958, in J. Taylor (ed.), *Selected Writings of John Hughlings Jackson*, Vol. 2, Staples Press, London.
20. Mercier, C.: 1925, 'Recollections', in J.H. Jackson, *Neurological Fragments* Oxford Univ. Press, Oxford.
21. Young, R.M.: 1970, *Mind, Brain and Adaptation in the Nineteenth Century*, Clarendon Press, Oxford.

SECTION II

PHILOSOPHICAL IMPLICATIONS OF PSYCHOSURGERY

JOSEPH MARGOLIS

PERSONS AND PSYCHOSURGERY

As every commentator appreciates, the controversy over psychosurgery depends largely on its implications for more comprehensive issues: the range of psychotechnology, the defensibility of behavioral control, the definition of illness and mental illness, the rights of patients and the obligations of physicians or the rights of society and the obligations of citizens and professionals, the tolerance of deviance or criminality or violence, the meaning of personal and social welfare, the demarcation of psychosurgery, the purpose of medicine, the conditions of rational cost-benefit policies, the nature of human values, the relation between psychological and behavioral phenomena and the physical condition of the human body. These are by no means wild associations; but to identify them suggests the ease with which debate quite naturally turns away from the strictest questions regarding psychosurgery and the need for constructing a unified account that would facilitate linking relevant issues with some fair accommodation of competing convictions.

In the narrowest sense of the current disputes, the foci of contention appear to include the following considerations: (a) whether psychosurgery is advocated for medical or non-medical benefits; (b) whether psychosurgery is, for either medical or non-medical benefits, advocated for diseased or non-diseased structures or tissue; (c) whether psychosurgery is properly so termed for all neurosurgery that yields significant psychological and behavioral changes or only for such surgery as is, within the purview of professional practice, designed to relieve mental illness, behavioral disorder, pain, or specifically socially deviant or socially undesirable behavior; (d) whether or not the state of the relevant disciplines is sufficiently advanced to justify technical expectations; (e) whether or not, given the state of the disciplines, the medical and non-medical risks of psychosurgery outweigh the anticipated benefits; (f) whether or not the actual practice of psychosurgery has, for various affected sub-populations, been conducted with due attention to the rights, particularly the informed consent, of patients,

S. F. Spicker and H. T. Engelhardt, Jr. (eds.), Philosophical Dimensions of the Neuro-Medical Sciences, 71–84.

professional peers, family, and the community at large as well as to the welfare and just and equal treatment of patients and others affected; (g) whether or not psychosurgery ought to be restricted to medical objectives only, constrained or not constrained in accord with certain legal, moral, professional, political, and economic safeguards, or ought to be construed as a specialty, instrumentally available for extramedical behavioral modification or control with or without similar safeguards; and (h) whether or not we now have or can reasonably develop procedures for appraising the psychological and behavioral consequences of specific psychosurgical operations.

Obviously, there are procedures other than psychosurgical, both medical and non-medical, for affecting, both permanently and temporarily, the psychological and behavioral characteristics of particular persons, possibly with as much or even more force than surgical procedures. Drugs and the chemical alteration of bodily processes, pollution and climate, psychoanalysis and indoctrination, training and education, operant conditioning and electroshock treatment, brain implantation, isolation and confinement, and trauma suggest the range of alternative possibilities – without, be it noted, any comparative assessment of the permanence, profundity, scope, specificity, or justifiability of such procedures.[1] Still, the nervous system and, in particular, the brain are so centrally involved in any realistic policy bearing on the freedom, welfare, well-being of individuals and human communities, and the practice and the enlargement of the practice of professional psychosurgery are, in principle, so readily accessible to rational control, that there is a special advantage and point to attempting to confine attention to psychosurgery itself. Also, the central literature is relatively compact:[2] there is very little reason to attempt another survey.

The general strategy in opposing psychosurgery may, not unfairly, be said to center on two claims: first, that surgery performed on such significant portions of the brain as the frontal lobes, the amygdala, the cingulum, the hypothalamus cannot but affect adversely a much larger range of psychological functioning than that normally specified in the justifying complaint (whether of medical dysfunction or of socially undesirable behavior); and second, that favorable findings with respect to such surgery normally fail to appraise consequences to the total personality and behavior of the patient apart from the original complaint or else construe such consequences, in accord with the personal convictions of the attending physician, solely or principally in terms of contributing to the correction of the original complaint. Dr. Peter R. Breggin is probably responsible for collecting the largest number of specimen instances against which the opposition strategy seems most reasonable.[3] But these two claims appear in themselves to be quite formidable general considerations counting, at least provisionally though for somewhat different reasons, against the entire practice of psychosurgery.

It is not unreasonable to say that we have only the barest sketch of the functioning of the brain, particularly of the cerebrum. And here, the most compelling findings seem to lead us in a variety of directions, not actually incompatible but certainly not neatly convergent. For example, we know that the cerebral hemispheres appear not to be structurally different in any significant respect and yet, normally but puzzlingly, hemispheric dominance obtains – usually, left-hemisphere dominance, associated with speech – together with the distinctly lateralized specialization of functions [2, 9]. Again, chiefly on the strength of the work of R.W. Sperry and his associates, we have been led to appreciate, surprisingly and somewhat against the global views of K.S. Lashley, the specificity of the neurological conditions of memory and learning (e.g., [24]). But also, in view of other evidence bearing both on the tolerance and the limits of tolerance of cortical ablation with respect to given functions, we are led to theorize along the lines of informational, that is functional, models of neural networks, for which some types of neurons and certain more complex structures may be particularly suited [23]. A further, most profound question regarding the brain is broached – but only broached – by Michael Gazzaniga: reflecting on early and late brain lesions and the puzzling phenomenon of callosal agenesis, Gazzaniga suggests that the "primary neurocircuits and organizations are laid down according to a specific set of instructions," "established by informational systems revealed through hereditary mechanisms," where "whether the initial capacity of the system is maximized depends upon environmental contingencies" ([9], ch. 8; [20]). Considerations of these sorts suggest the extraordinary difficulty involved in justifying ablations of the brain of such significance as is at last promisingly linked with the complaints to be relieved. This, for at least two reasons: first, because, on the evidence, any structure of the brain associated with memory, thought, emotion, and behavior is bound to be linked causally with a very complex set of psychological states and dispositions of much greater scope than any complaint normally introduced to justify psychosurgery could be expected to specify; and second, because the complaint itself – if not fairly narrowly construed in medical terms (as for instance in relieving epilepsy) – is primarily focused on what may be termed behavioral syndromes (for instance, criminality, homosexuality, aggressiveness, hyperactivity) or associated psychological syndromes (for instance, homosexual fantasy, persistent hallucination, "over-subtle reasoning," neurotic mentation, "excessive self-consciousness") [3] and not on particular structures of the brain that quite usually are not diseased or dysfunctional within the ordinary terms of physical medicine.

In fact, on the most promising theories, complaints of the sort given may be expected to be focused primarily on the informational or functional circuits of the brain rather than on the specific structural parts of the brain itself ([25], esp. pp. 264-266). Certainly we see here the practical implications of questioning classical mind/brain identity theories – for instance the well-known view of J.J.C. Smart [21] – without yet deciding whether or not an alternative identity theory (identifying the mental and the functional) is required or is most reasonable.[4] Even Sperry's experiments show that certain specialized patterns of learning and memory can be eliminated by ablation, *not* that ablation can be very narrowly localized with respect to psychological function or behavior. The result is that psychosurgery seems more closely linked with the elimination of a general *repertory* (including addition, by learning, to such a repertory of psychological and behavioral elements) than with the alteration, suppression, or elimination of narrowly determinate *elements* within such a repertory. Hence, the justification of psychosurgery seems to be conceptually bound, even under the most informed conditions, to pursue some sort of cost-benefit comparison of the by-consequences and the intended relief. Also, emphasis on the functional rather than the structural features of the brain at least favors the presumption that the brain functions as a unified system. The upshot is that the effect of even the most advanced surgery is substantially affected by the total personality of the patient – which bears in an important way on the predictable outcome of any given surgery as opposed to the merely predictable reduction of severe brain deficit [25]. But this leads us on to the second claim of the suggested counterstrategy. In any case, the defense of psychosurgery seems often implicitly to proceed by a sort of Principle of Double Effect ([7], pp. 1020-1022); but defenses of that sort are rather dubious when it is clear that the foreseen consequence is normally unavoidable and that, therefore, the distinction between intended and foreseen consequences strikes us as merely contrived.[5]

Here the definition of psychosurgery is critical, because it is obvious that competing characterizations, however circumscribed their scope, betray the policy convictions of their respective advocates. A few examples will suffice. Vernon Mark says: "In its classical sense, psychosurgery includes operations on the frontal lobes or their connections in the brain to relieve the symptoms of intractable depression, agitation, compulsion, delusion, hallucination and ideas of reference in patients with no known brain disease"; also, that "Psychosurgery has expanded beyond its classical meaning to include any neurosurgical operation that affects human behavior, even if the patients being treated have obvious brain disease... most controversially patients... who have epilepsy and psychiatric symptoms, often including abnormally aggressive behavior" ([16] pp. 217-218). Mark, a neurosurgeon himself, restricts operations to medically (at least psychiatrically) specified complaints, stresses 'intractable' and 'abnormally' extreme psychological and behavioral symptoms, and conveys a temperate view on candidacy for psychosurgery. Stephan Chorover, a specialist in brain physiology, holds that "psychosurgery may be defined as brain surgery that has as its primary purpose the alteration of thoughts, social behavior patterns, personality characteristics, emotional reactions, or some similar aspects of subjective experience in human beings," though he concedes that its proponents "make the more specific claim that with respect to certain forms of mental illness, behavior disorder, or emotional disturbance, significant therapeutic results may be obtained following the surgical destruction of particular brain regions."[6] Chorover, however, is persuaded that the categories of mental illness and behavioral disorder are being pressed by the new psychotechnology into the service of "essentially totalitarian" objectives ([4], pp. 43-49, 52, 54, esp. 44). George Annas and Leonard Glantz, specialists in the legal aspects of medicine, define psychosurgery in a way that is utterly neutral to both medical and legal concerns, in order precisely to permit a medical or legal review of practice (especially malpractice) to be undertaken: they restrict psychosurgery to "any procedure that destroys brain tissue for the primary purpose of modifying behavior" ([1], p. 249). Robert Neville, a philosopher, concludes that, "for scientific purposes... psychosurgery should be understood as including all neurosurgery that has direct psychological effects." He accepts the extended scope of the term because, although he himself stresses the usually favored "difference between operating on a clearly damaged brain that has pathological symptoms and operating on a brain that is apparently normal to cure pathological behavior whose sources are apparently environmental and whose form, though intractable to other treatments, is learned," he notes that "the acceptance of this dichotomy might give psychosurgeons license to perform whatever procedures they desire on brains in which they can find some abnormality" ([17], p. 346).

The tendentiousness of all of these maneuvers is plain. Construe psychosurgery as a medical procedure: apart from ulterior ethical and professional considerations, the legitimacy of medicine entails the legitimacy of psychosurgery – coping with 'intractable' aggressiveness or hyperactivity is rather like removing a brain tumor. Construe it as a neutral medical procedure: the serious question that arises is whether, in employing it, a practicing surgeon has opened himself to a malpractice suit or whether the rights of patients and other interested parties ought to be protected in a fresh way – for instance, against certain forms of medical experimentation. Construe it as a procedure for altering and controlling the mental life of human beings: it becomes primarily a question of ethical and political policy, by reference to which the very practice of medicine itself must be governed – for instance, to protect us against Big Brother. Construe it in the broadest neurosurgical terms: an ethical review of medical practice is at least insured – though the guidelines for such review may themselves be open to considerable quarrel. But though they are all partisan proposals, there is no 'objective' alternative. Ultimately, what this means is that, whatever the irreducibly minimal considerations of medicine and ethical review may be, any and every full conception of health and disease and any and every full conception of the well-being and welfare of men must be ideologically qualified [10, 11, 12, 14, 15].

Here, it becomes difficult to sort out the essential considerations in an orderly way. Two strategic qualifications, however, may lead us rather undisputatiously into the eye of the storm. First of all, functional disorders, on any plausible view and however generously construed, cannot be said to entail structural disorders; and where psychological or behavioral 'disorders' are said to obtain (hyperactivity, for instance, or aggressiveness), such disorders should not, for that reason alone, be identified as functional disorders of the brain or nervous system (as 'minimal brain dysfunction,' for instance), that is, suitably qualified without independent evidence of such dysfunction ([4], p. 49). These qualifications remind us that calling something a medically significant disorder or disease does not make it such, and that, whatever the conceptual defense of the categories of physical disease, the defense of the categories of mental and behavioral illness are best viewed as an extension of the competence of medicine and are notoriously open to doctrinal and ideological manipulation [10]. Second, the admission that psychosurgery may effectively relieve a complaint does not entail that it was caused by some functional or structural disorder and entails nothing regarding the classification of that complaint in medical or ethical or legal terms or regarding its comparative importance in any schedule of medical, ethical, or legal concerns. These qualifications remind us that etiology and therapy are only contingently linked and that the conceptual defense of the normative categories of medicine, ethics, and the law is by no means obvious.

Now, the single most important theme that appears congruently in medical and ethical contexts concerns the relationship between individual members of a population and that population. Determinations of health and disease, for example, are normally specified for individual organisms, but the justified provision of a tolerated range of normality or normal functioning makes no sense apart from the survival and the conditions of survival of an entire population [15]. According to the balance theory of population structure, for instance, advocated by Theodosius Dobzhansky,

> Normal, highly adapted populations contain some genes held in a state of balanced polymorphism. The genetic loads that populations carry are, then, of at least two kinds. A mutational (classical) load consists of genes that damage heterozygotes as well as homozygotes, or if only the latter then at least do not benefit the heterozygotes. A balanced load consists of genes and gene combinations which are advantageous in heterozygotes as compared to their effects in homozygotes. Balanced loads are maintained chiefly by natural selection rather than by recurrent mutation" ([6], pp. 296-297).

By 'polymorphic,' Dobzhansky means the feature of a species in virtue of which different sub-populations are genetically variable, containing as many genotypes as individuals (identical twins excepted) ([6], p. 221). Thus, the gene for sickle cell anemia, in man, is thought to be almost completely lethal for homozygotes, relatively benign for heterozygotes, *and* particularly conducive to the fitness of a *population* exposed to malaria (conveying immunity to heterozygotes) – where malaria is normally lethal for 'normal' homozygotes (for instance, among the blacks of Central Africa).[7] This shows very nicely the conceptual dependence of what will pass as the boundaries of health and disease in individuals, and on what will pass as the boundaries of health and disease in the aggregated populations of a species. Similarly, in ethical contexts, it is extremely difficult (though, arguably, not impossible) to formulate the conditions for an ehtically correct or good life for individuals in a way that is not conceptually dependent on conditions thought to be required for an entire population. At the present time, for instance, disputes about the adequacy of utilitarianism and theories of justice are quite explicit about this [19, 22]. The adequacy of such accounts is not the issue; nor is the issue merely one of universalizability, which is in any case a matter of consistency only ([12], pp. 84-87). It is rather that ethical concerns are, at least minimally, concerns regarding how the behavior of particular persons affects the interests of others: rational planning of any sort with respect to such contingencies requires reflection on the acceptable level of life of all those affected, including tolerable ranges of deviance and normality.

But the convergence between medicine and ethics in this respect holds deeper implications – which, it can be shown, bear in a decisive way on the justification of psychosurgical policies. For one thing, the concept of the normal functioning of an organism – whether man or lower animals or even plants – is most reasonably construed in terms of certain statistically dominant prudential interests (or analogical extensions of such interests to plants and lower animals), for instance, survival, control of pain, gratification of desire, capacity or power to use one's body to fulfil characteristic enterprises. The list is informal and need be no more than that, open to extension in other contexts (for instance, in terms of security of various sorts and access to material goods).[8] Its dialectical advantage is, precisely, that it obviates the need to attribute an essential normative function to human beings – which in any case is difficult to defend and is bound to be tendentious[9] – and that it provides a basis for the conceptual unity of such diverse disciplines as medicine, the law, politics, ecology, ethics, and economics. But, in doing so, it also signifies that, in the absence of any defensible theory of natural function, the determinate forms of prudential interests and ulterior normative objectives countenanced in medical and ethical settings are themselves expressions of ideologies and normative convictions prevailing, however conservatively and plausibly, in one society or another. Human life falls between birth and death: hence, neither death nor aging as such are construed as medical disorders or as ethically relevant evils. Our expectations about what a normal lifespan is and what level of metabolic vitality and capacity constitutes the absence of disease will be a function of how our prudential interests are thought determinately to facilitate our legitimate ulterior interests ([14], ch. 7, [15]). Obesity, for example, will obviously be appraised differently, with respect to the specification of disease, in terms of life expectancies that are themselves a function of technological achievements and cultural expectations relative to such achievements. Similarly, no significant range of human activities can fail to bear directly on the prudential interests of others, but precisely how determinately affecting the survival and the quality of life of others will be ethically appraised is clearly a function of the ethical traditions prevailing in one society or another.

Psychosurgery, however, as the usual definitions canvassed make clear, is itself construed as a medical procedure, indicated where brain damage or brain dysfunction obtains (as in brain tumors or epilepsy) or where, in the absence of physical disease or dysfunction, psychiatrically qualified disorders obtain (as in intractable depression and delusion) or where, in the absence of disease or medical disorder of any sort, socially deviant or deficient or unacceptable behavior obtains (as in aggressiveness, violence, criminality, homosexuality, nonconformity, hyperactivity). The problems here are rather complex because (1) even the least tendentious array of purely physical diseases is still conceptually controlled by our theories of how health is specified (relative to minimal prudential concerns and prevailing social expectations); (2) mental illness and behavioral disorder, however conservatively linked with the prudential concerns of physical medicine, are bound to incline markedly toward the ideological values of a society;[10] and (3) because the political, legal, and ethical values of a society, admitting their congruence with minimal prudential interests, are even more prone to ideological direction than the norms of somatic and psychiatric medicine. So seen, there is not as great a difference at stake in debating whether psychosurgery should be applied in putatively psychiatric cases where brain disease is absent or in cases that are frankly viewed in terms of the reduction or elimination of socially undesirable mentation and behavior. Preposterous findings have been reached in both contexts [3].

We are now in a position to locate the general strategy opposing psychosurgery in a more ramified context than before. Significant psychosurgery appears to be incapable – not only in practice but in principle – of suppressing or extirpating quite specific elements of mentation and behavior without also substantially affecting, precisely where successful, larger ranges of the psychological and behavioral repertories of particular patients; *and* the repertories affected on any plausible view, including those elements involved in the original complaint, are just the ones most sensitive to doctrinally variable psychiatric and ethical appraisal. In particular, complaints may well focus on episodes characterized in terms of a society's legal or other appraisive categories, whereas surgery affects portions of the physical brain, the functional role of which cannot, in principle, be restricted to the terms of such adverse judgments. Hence, at the very least, a distinctly conservative view seems reasonable where there is doubt about the objective validity of psychiatric and ethical norms or, more, even about the prospect of discovering such norms – particularly in a political context that favors, within limits, pluralistic norms of a personal and social nature. Nevertheless, it must be said, in all fairness, that such an argument is dialectical rather than categorical, in the sense in which it assesses the use of the procedure in terms of its conformity, in a favorable or unfavorable degree, with ideological currents prevailing, at the present time, in our own society. It can be strengthened in the direction of a categorical rejection if, (a) construed as a medical procedure, the coarseness of its techniques regularly affects adversely whatever is relatively uncontroversially required in order to be free of disease [15]; or (b), construed in terms of ethical, legal, and political values, it tends to disable (as in restricting the repertory of responses or the capacity to learn new responses) just those psychological abilities (cognitive, conative, affective) that play a central role in the prudential concerns of the species, which our ethical, legal, and political values must themselves elaborate in ideologically determinate ways. But it is possible, in principle (and, on the evidence, not entirely unlikely), to formulate the justifying complaint in such a way that it conforms to the coarseness of the available techniques.

On the other hand, the life of man *is* ideologically organized. Conceptions of personal health and disease are inevitably connected with the survival and activity of an entire society; and personal liberty and well-being are, equally, functions of the perceived organizational needs of an entire society. Moreover, questions of health and well-being concern human persons, not merely members of *Homo sapiens*; but persons, as opposed to human animals, are at the very least culturally emergent entities – insofar as they exhibit linguistic mastery and the capacity of self-reference.[11] But if that is so, then, (1) as culturally emergent, persons cannot be expected to have determinate natural functions in virtue of which ethical norms may be objectively laid down; (2) precisely in emerging as culturally groomed creatures, they must be ideologically indoctrinated; and (3) in the pursuit of their own favored way of life, societies cannot but impose constraints construed (both in medicine and in ethics) in terms of the self-interest of those affected and in terms of the interests of the society itself [14]. Hence, assuming that psychosurgery may be practised in a way that cannot be (more or less) categorically condemned, the only alternative strategy is dialectical. In that case, particularly in a political context in which radically divergent doctrines are tolerated and not disallowed, it is not in the least improbable that a quite coherent defense of psychosurgical procedures (waiving here all questions of the comparative merits of alternative forms of brain modification and control) may be made out. But that would merely confirm what we may have suspected to begin with: that we cannot escape being partisans on the issue and that the best we can expect is to be rational partisans. In any case, there is no 'natural' human person to be protected from cultural 'intrusions' – whether by surgery, electrodes, or education – only culturally emergent entities groomed from pre-cultural potentials, about the manipulation and treatment of which or whom we have remarkably different convictions [5].

A final word may be helpful. The issues raised by psychosurgery are hardly peculiar to itself, and the effort to move debate from medical to ethical issues does not insure a gain in clarity or power. The concepts of rights, of what is right or wrong, and of health and disease depend on essentially the same speculations; and the position here sketched is intended to fall between the extremes of extreme skepticism or relativism and extreme objectivism or absolutism – both with respect to medical and with respect to ethical concerns.

Temple University,
Philadelphia, Pennsylvania

NOTES

[1] José Delgado [5] identifies the principal types of brain control as "(1) mechanical, through surgery, (2) electrical, through brain stimulation, and (3) chemical, by placing drugs inside the brain... then, there is a fourth type of control, the effect of the environment on the electrical and chemical activity of the brain," p. 2.

[2] A fair sample of the principal views may be had from the following: 1974, 'Symposium: Psychosurgery', *Boston University Law Review* **54**, 215-355: Packet #900, 'Behavioral Control: Psychosurgery and Electrical Stimulation of the Brain', *Readings in Society, Ethics and the Life Sciences*, distributed by the Institute of Society, Ethics and the Life Sciences; and W. Gaylin, J. Meister, and R. Neville (eds.): 1974, *Operating on the Mind*, Basic Books, New York. What is not included in these collections is almost certainly cited. In addition, three extremely useful books – one a survey, the others collections – are Valenstein, E.S.: 1973, *Brain Control*, John Wiley, New York; Szasz, T.A. (ed.): 1973, *The Age of Madness*, Anchor, New York; and London, P.: 1969, *Behavior Control*, Harper and Row, New York.

[3] Breggin [3] is entirely candid about his very strong opposition to psychosurgery – which some have thought led him to neglect some serious supporting evidence; but certainly his specimens are extraordinary. For example, he reports the hypothalamic surgery of a certain F.D. Roeder, in West Germany, "in an effort to cure 'sexual deviation,' "; Roeder reports as his accomplishment, "Potency was weakened, but preserved.... The aberrant sexuality of this patient was considerably suppressed, without serious side-effect. One important feature was the patient's incapacity of indulging in erotic fancies and stimulating visions..." ([3], p. 458). There are a number of such instances that Breggin cites, and there seems to be little reason to challenge the accuracy of his findings, despite his partisan convictions. *Vide*, also, Chorover, S.L.: 1974, 'Psychosurgery: A Neuropsychological Perspective', *Boston University Law Review* **54**, 231-248; and Valenstein, E.S. (n. 2).

[4] Cf. Fodor, J.: 1968, *Psychological Explanation*, Random House, New York; Putnam, H.: 1960, 'Minds and Machines', in S. Hook (ed.), *Dimensions of Mind*, New York University Press, New York; Putnam, H.: 1967, 'The Nature of Mental States', in W.H. Capitan and D.D. Merrill (eds.), *Art, Mind, and Religion*, University of Pittsburgh Press, Pittsburgh; Margolis, J.: 1973, 'The Perils of Physicalism', *Mind* **82**, 566-578. Some have even tried to combine both sorts of identity theory; cf. Lewis, D.K.: 1966, 'An Argument for the Identity Theory', *Journal of Philosophy* **63**, 17-25; and 1972-1973, 'Psycho-physical and Theoretical Identification', *Australasian Journal of Philosophy* **50**, 249-258.

[5] *Vide* Anscombe, G.E.M.: 1961, 'War and Murder', in W. Stein (ed.), *Nuclear Weapons: A Catholic Response*, Sheed and Ward, New York – where the issue is discussed in another context; see also ([14], chs. 3-4) regarding constraints of the Principle.

[6] Chorover, S.L.: 1974, 'Psychosurgery: A Neuropsychological Perspective', *Boston University Law Review* **54**, 231. He is here reporting the view also of Dr. Bertram Brown, Director of the National Institute of Mental Health. Cf. Oregon Laws 1973, chs. 615, 616.

[7] ([6], pp. 150-152). Dobzhansky refers particularly to the work of A.C. Allison.

[8] A fuller account of prudential interests and their systematic role in all normative inquiries is provided in [14].

[9] The classical form of the the theory that there are natural norms of human nature appears in the eudaimonism of Plato and Aristotle. A version of the most general schema of such a view is given in Hampshire, S.: 1959, *Thought and Action*, Chatto and Windus, London, p. 223; but it can be shown to be untenable. Cf. [12], ch. 5.

[10] The most active critic of the medical interpretation of mental illness is undoubtedly Thomas A. Szasz; *vide* 1961, *The Myth of Mental Illness*, Harper-Hoeber, New York; 1970, *The Manufacture of Madness*, Harper and Row, New York; but also [10] for a critical review of Szasz's position.

[11] The details of this intriguing problem – distinguishing between human bodies, human animals, and human persons – are not essential here, only the fact that persons cannot be identified as such except in a cultural context. A hint of the account here favored appears in 1974, 'Works of Art as Physically Embodied and Culturally Emergent Entities', *British Journal of Aesthetics* **14**, 187-196; a full account is provided in an as yet unpublished book, *Persons and Minds*.

BIBLIOGRAPHY

1. Annas, G.J. and Glantz, L.H.: 1974, 'Psychosurgery: The Law's Response', *Boston University Law Review* **54**, 249-267.
2. Bogen, J.E.: 1969, 'The Other Side of the Brain I: Dysgraphia and Dyscopia Following Cerebral Commisurotomy', *Bull. Los Angeles Neurol. Soc.* **34**, 73-105; 1969, 'The Other Side of the Brain II: An Appositional Mind', *Bull. Los Angeles Neurol. Soc.* **34**, 135-162; Bogen, J.E. and Bogen, G.: 1969, 'The Other Side of the Brain III: The Corpus Callosum and Creativity', *Bull. Los Angeles Neurol. Soc.* **34**, 191-220; and Bogen, J.E., R. De Zure, W.D. Tenhouten and J.F. Marsh: 1972, 'The Other Side of the Brain IV: The A/P Ratio', *Bull. Los Angeles Neurol. Soc.* **37**, 49-61.
3. Breggin, P.: 1973, 'The Return of Lobectomy and Psychosurgery', reprinted from *Quality of Health Care – Human Experimentation*, in Hearings before the Subcommittee on Health of the Committee on Labor and Public Welfare, United States Senate, Ninety-third Congress, First Session, on S. 974, S. 878 and S.J. Res. 71, February 23 and March 6, 1973, Part 2, pp. 455-477.
4. Chorover, S.L.: 1973, 'Big Brother and Psychotechnology', *Psychology Today* 7, 43-54.
5. Delgado, J.: May 1973, 'Physical Manipulation of the Brain', *The Hastings Center Report*, Special Supplement.
6. Dobzhansky, T.: 1962, *Mankind Evolving*, Yale University Press, New Haven.
7. 'Double Effect, Principle of', *New Catholic Encyclopedia*, Vol. 4, McGraw-Hill, New York.
8. Fodor, J.: 1968, *Psychological Explanation*, Random House, New York.
9. Gazzaniga, M.S.: 1970, *The Bisected Brain*, Appleton-Century-Crofts, New York.
10. Margolis, J.: 1966, *Psychotherapy and Morality*, Random House, New York.
11. Margolis, J.: 1969, 'Illness and Medical Values', *The Philosophy Forum* **8**, 55-76.
12. Margolis, J.: 1971, *Values and Conduct*, Clarendon Press, Oxford.

13. Margolis, J.: 1973, 'The Perils of Physicalism', *Mind* **82**, 566-578.
14. Margolis, J.: 1975, *Negativities. The Limits of Life*, Charles Merrill, Columbus.
15. Margolis, J.: 1976 (forthcoming), 'The Concept of Disease', *Journal of Medicine and Philosophy* **1**.
16. Mark, V.H.: 1974, 'Psychosurgery versus Anti-Psychiatry', *Boston University Law Review* **54**, 217-230.
17. Neville, R.: 1974, 'Pots and Black Kettles: A Philosopher's Perspective on Psychosurgery', *Boston University Law Review* **54**, 340-353.
18. Putnam, H.: 1960, 'Minds and Machines', in S. Hook (ed.), *Dimensions of Mind*, New York Univ. Press, New York; 1967, 'The Nature of Mental States', in W.H. Capitan and D.D. Merrill (eds.), *Art, Mind, and Religion*, University of Pittsburgh Press, Pittsburgh.
19. Rawls, J.: 1971, *A Theory of Justice*, Harvard University Press, Cambridge.
20. Rose, S.: 1973, *The Conscious Brain*, Alfred A. Knopf, New York, esp. ch. 8.
21. Smart, J.J.C.: 1962, 'Sensations and Brain Processes', revised in V.C. Chappell (ed.), *The Philosophy of Mind*, Prentice-Hall, Englewood Cliffs.
22. Smart, J.J.C. and Williams, B.: 1973, *Utilitarianism, For and Against*, Cambridge University Press, Cambridge.
23. Sommerhoff, G.: 1974, *Logic of the Living Brain*, John Wiley, London, esp. chs. 7-8.
24. Sperry, R.W.: 1947, 'Cerebral Regulation of Motor Coordination in Monkeys Following Multiple Transaction of Sensorimotor Cortex', *Journal of Neurophysiology* **10**, 275-294; 1968, 'Plasticity in Neural Maturation', *Developmental Biology Supplement* **2**, 306-317; 1966, 'Brain Bisection and Mechanisms of Consciousness', in J.C. Eccles (ed.), *Brain and Conscious Experience*, Springer-Verlag, New York.
25. Valenstein, E.S.: 1973, *Brain Control*, John Wiley, New York.

JERRY A. FODOR

PSYCHOSURGERY: WHAT'S THE ISSUE?

I

I had better begin my commentary on Professor Margolis' paper with a number of disclaimers. I do not have any very extensive knowledge of the polemical literature on psychosurgery, and such philosophical and psychological background as I have does not seem to me to bear very closely on the question. I therefore find myself approaching the topic with no noticeable advantages over any other reasonably reflective layman. No doubt, it is for this reason that I seem to be having some difficulty in understanding what the topic *is*. That is, it is not clear to me what issues of principle are supposed to be involved in psychosurgical procedures that are not involved in other, less dramatic, medical practices. Since it may be that there are others who share my difficulty, I thought what I might do is simply to enumerate some of the questions that might be candidates for *the* issue about psychosurgery and suggest why these candidates, at least, strike me as unlikely to get elected.

II. THE PRACTICAL ISSUE

One clearly pertinent question, that is clearly so identified in Professor Margolis' paper, is this: do we know enough about the way the brain works to rule out the practical likelihood of adventitious side effects of psychosurgical procedures? This is, in at least two respects, not a question of *principle*. First, to the extent that the answer may be 'no,' the most that that could show is that it is unadvisable to employ such procedures *now*; viz., in our current state of ignorance. It would, in particular, say nothing about the admissibility of employing such procedures in some conceivable future state of medical knowledge. Second, the possibility of adventitious side effects would have to be weighed against the likelihood of positive gains

S. F. Spicker and H. T. Engelhardt, Jr. (eds.), Philosophical Dimensions of the Neuro-Medical Sciences, 85–94.

in any rational evaluation of the treatment. So, even if such side effects are very likely to occur, there might, at least in principle, still be circumstances in which the use of the treatment would be warranted, assuming, as we are now doing, that the possibility of side effects is the only argument against the treatment.

I say that the present question is not, so far as I can see, a question of principle; which is not, of course, to claim that it is not an important question or one which may properly determine how we ought to act. I am, so far as I can tell, entirely in agreement with Professor Margolis' intuitions here. In fact, I am inclined to put the case even more strongly than he does. It seems to me to be strictly correct to say that, as things now stand, nobody has the slightest idea of how the brain works. I am not, let me remark, committing the folly of denying that a great deal of neuroanatomy and neurophysiology has been successfully worked out; nor am I unappreciative of the profundity of some of the inquiries in this area. What I am denying is that there exist any serious and general accounts of the *functional* organization of the brain; specifically, of how it produces its psychological effects. This seems to me to be the case at every level of analysis, from the most abstract to the most detailed. In the former case, we do not *even* know whether the brain is best viewed as a digital or an analog information processing system. (Indeed, the very question whether the brain is, in any substantive sense, properly to be viewed as an information processing system *at all* is, I think, plausibly considered moot.) At the level of extreme detail, we do not even know what neurophysiological entities are the functional units in terms of which the psychological capacities of the brain are organized. And, so far as I can see, the situation is no better in the middle ranges. It is, I take it, not a tendentious claim but a mere platitude that we are not within miles of an adequate account of the neurophysiology of memory, or perception, or thinking, or talking, or acting. What we do have, and what I suppose any practical psychosurgery would have to be based upon, is some pretty crude and patently *post facto* generalizations about the observable effects of specified kinds of intervention. So far as I can see, that is *all* we have.

I suppose that medical practitioners have often found themselves in this position: one has to act (or abstain from acting) on the basis of empirical generalizations about past outcomes rather than on a secure physiological theory. Even so, the psychosurgical case seems to me to be unusually bad.

For, as Professor Margolis also points out, our ability to *measure* the effects of psychosurgical interventions is no better than the construct validity of our psychological tests, and the construct validity of available psychological tests is notably dubious. Suppose, for example, that someone were to invent a psychosurgical procedure for which he claimed some clinical benefit (e.g., the reduction of disabling anxiety) without impairment of cognitive functioning. How could we tell whether his claim is *true*? Clearly, we could do so only if we had reliable tests of cognitive functioning. But, equally clearly, the reliability of our tests of cognitive functioning is unlikely to be better than the theories of cognition that the tests are based upon, for it is the theories which formulate such insight as we have about what cognitive functioning *is*. And clearest of all is that our current psychology offers no account of cognitive functioning that it is possible to take seriously.

I do not really suppose that there could be much argument about the extent of our inability to measure the consequences of psychosurgical intervention, so I shall not dwell on the point. Let me just remind you of one particularly spectacular example; Until quite recently, neurophysiologists had no very clear idea of what the interhemispheric fibers are for. (I have heard Lashley quoted as suggesting that they exist solely to transmit epilepsy from one side of the brain to the other, but I do not vouch for the accuracy of the attribution.) The other side of the same historical fact is that, until recently, no one was able to find any very striking consequences of sectioning these fibers. Times change, however, and it is now fashionable to claim that the interhemispheric fibers hold the mind together, and that sectioning them produces two persons resident in the same body. If true, that is a side effect on a fairly cosmic scale. I see no reason to suppose that there are not other equally appalling discoveries to be made about the consequences of other psychosurgical interventions.

I said this was a practical point, and the practical consequences would seem to be clear enough. All other things being equal, the advisability of a treatment is surely inversely related to the probability of unintended side effects. And, all other things being equal, the probability of such side effects is surely proportional to our ignorance of the functioning of the affected organ. In the case of brain function, our ignorance is as near total as makes no difference. The rational course of action would therefore be to follow as conservative a course as the circumstances will allow.

III. ARE BEHAVIORAL DISORDERS DISEASES?

If I understand Professor Margolis' paper correctly, his view of the issue about psychosurgery is roughly this: in the contentious cases, psychosurgery is directed at the treatment of behavioral disorders. Now, *if* behavioral disorders are disease conditions, there is anyhow a *prima facie* case for taking them to be species of neuropathologies. And *if* they are species of neuropathologies, there is at least a *prima facie* case for taking psychosurgical treatment as raising, as it were, *only* internal medical issues. Professor Margolis is, I take it, suspicious of this line of argument, primarily on the ground that he takes questions of what count as health and disease as ineluctably moral and, indeed, ideological. "...in the absence of any defensible theory of natural function, the determinate forms of prudential interests and ulterior normative objectives countenanced in medical and ethical settings are themselves expressions of ideologies and normative convictions prevailing, however conservatively and plausibly, in one society or another." Now, I do not have any quarrel with this,[1] but it does seem to me to prove too much. For, suppose that questions of disease and health *are* moral and ideological to any extent you like. So long as this is true of *all* questions of health and disease, this consideration does not provide for a principled distinction between what is at issue in psychosurgery and what is at issue in, say, dental surgery. I suppose what Professor Margolis objects to is a kind of medical Positivism which goes: it is simply a matter of fact whether, say, Smith is neurotic; and, if he is, it is simply another matter of fact whether cutting out his cerebellum is a way of curing him. Professor Margolis wants, understandably, to remind us of the moral issues that (may) have been begged by terms like 'neurotic' and 'cure.' But his way of doing so strikes me as slightly curious. If all questions about pathology are finally ideological, then so are all questions about dental caries. And, I should have thought, the Positivist will be satisfied if the moral issues involved in psychosurgery are continuous with those involved in having your teeth drilled.

And, anyhow, it is hard to believe that the question of principle *is* the question whether behavioral disorders are diseases. For, even aside from the sorts of practical issues raised above, their being so would surely not be either necessary or sufficient for psychosurgery to be warranted. It is not necessary because psychosurgical procedures might easily be warranted in *non*-pathological conditions. If, for example, it were possible to make people smarter by psychosurgical intervention or to make them live longer, then I suppose such intervention would be morally permissible *given*, of course, that they were voluntarily submitted to. And, if in principle it is morally permissible to use psychosurgical procedures to make people smarter, it is not clear to me why it should not, in principle, be morally permissible to use such procedures to make them happier or better. After all, we do not think cosmetic surgery is immoral (even though we may think it is silly). And cosmetic surgery does not even *purport* to cure a disease. I do think these points raise issues about one's rights over one's own body, and I think such issues are implicated in the psychosurgery dispute, as in other questions of medical practice. I will return to this presently.

What I have argued so far is that it is not a *necessary* condition for the justification of psychosurgery that the conditions it purports to treat should be diseases. My next point is that it is not *sufficient* either. I think this should be clear from untendentious cases. Consider, once again, my dental caries. A tooth with a cavity is, I suppose, a clear case of a tooth in a diseased condition. And I can admit, for purposes of this discussion, that, all things being equal, it is better for a tooth (any tooth) to be free of cavities than for it to be ridden. Still, it does not follow from the fact that my tooth has a cavity that it would be all right for someone (you, the American Dental Association, the Federal government, or anyone else) to fill my tooth. I suppose it would be all right if (but *only* if) it is all right *with me*. If it is *not* all right with me (if, for example, I make a hobby of collecting cavities and want the ones I have collected left alone), then there is no one who has a right to take my cavity away, and this despite the (presumed) fact that cavities are diseases and I would be better off without the ones I have got.

What is at issue here is a very general principle: From the fact that S is a state of affairs it would be desirable to bring about, it does *not* follow that there is someone (you, me, the American Dental Association, the Federal government, or anybody else) who has a right to act so as to make S the case. So, in particular, even if all behavioral disorders are diseases, and even if, all things being equal, diseases are best alleviated, it does not follow that anyone has the right to take such steps as might be required to alleviate the behavioral disorders. *A fortiori* it does not follow that anyone has the right to take such psychosurgical steps as might be required.

It should be noted, by the way, that just as the possibility that my reasons for not wanting to be treated may be bad reasons does not impugn my right not to be treated if I do not want to be, so the fact that your motives for wanting to treat me may be unimpeachable does not, in and of itself, bestow on you a right to undertake the treatment. Questions of motives are, in fact, simply not in place here. I mention this because it is sometimes suggested in discussions of psychosurgery that the real issue is somehow the purity of intentions of the practitioner. Of course, if it could be shown that psychosurgery is something that Wall Street cooked up to keep the workers in line (or, for that matter, something the workers cooked up to keep Wall Street in line) that would be interesting because it would suggest that we ought not place much credence in claims for the beneficial results of psychosurgery, and that would make the sorts of practical issues raised above even *more* pressing than I have taken them to be. But, however this may be, the present point is that it is *not* relevant to the issue of moral warrant. In particular, the fact that one intends to bring about a desirable state of affairs by doing such and such a thing does not, in and of itself, give one the right to do that thing; for that matter, the fact that one intends to bring about an *un*desirable state of affairs by doing such and such a thing does not, in and of itself, abrogate such right to do that thing as one may happen to have.

To summarize, people have rights *vis-à-vis* their own bodies, one of which is the right not to take, or to permit to be taken, such steps as would be required to alleviate its pathologies. This is not, of course, to deny that there may be limits to the exercise of this right, and in some (though certainly not all) controversies involving psychosurgery, the location of these limits may be what is at issue. So, imagine a world where dental cavities are catching. One would, I suppose, have a right to one's cavities even in that world; but though one would have a right to keep them, one would not have a right to *spread* them. And it is easy to imagine circumstances in which the only way to prevent your spreading them might be to cure you of them. I suppose the same sort of point applies, *mutatis mutandis*, to behavioral disorders; some such (though, obviously, not all such) might have consequences that infringe on other people's rights, and the circumstances might be that the only way of protecting those rights is to cure the disorders. But notice that one undertakes a considerable burden if one undertakes to show that such a case obtains. Roughly, one must show that limiting a person's rights over his body is the best way (or, anyhow, the best way available) of dealing with the case. In particular, one needs to show that there is no other way of dealing with it that does less damage to the rights and interests of the persons involved. Perhaps it goes without saying that to show this you need to prove quite a lot more than that mass ablation would be socially convenient.

Where we have gotten to is this: the moral justification of psychosurgery does not turn on whether the behaviors it purports to treat are, in fact, diseases. Psychosurgery might be justifiable even if they are not, and it might be unjustifiable even if they are. What does seem to be concerned is the rights that people have: rights to do things, or have things done, to their bodies if they wish to, and not to do them and not to have them done, if they wish not to. But I do not suppose that there is anything here that is *peculiar* to the discussion of psychosurgery. My rights *vis-à-vis* my frontal lobes seem, so far at least, to be all of a piece with my rights *vis-à-vis* my left ear. I have, or so it seems to me, a right to contract for the removal of either, and neither may be removed in violation of my rights.

IV. CONSENT

My rights *vis-à-vis* my body are, roughly, rights which imply that no one may act upon it without my consent. Now, notoriously, rights that involve consent tend to be hard to specify, and it might be argued that it is here that the special problems about psychosurgery arise. For, one might argue either (a) that it is intrinsic to the kinds of disorders that psychosurgery is aimed at that people who suffer from them must *ipso facto* be incompetent to give (or withhold) rational consent, or (b) that there is something so special about the conditions that psychosurgery aims to treat that it ought not be practised even upon subjects who consent to (or demand) it. I do not myself see much plausibility to such claims, however; not, at least, stated that baldly.

(a) There might be cases in which the very fact that psychosurgery is the treatment of preference makes it implausible that the subject is competent to give or to withhold consent. If, for example, psychosurgery were indicated in cases of extreme psychosis, we would perhaps have such an instance. I have only a couple of remarks to make about this. First, that not all (and, indeed, not the most contentious) cases are like this. The cases people are most worried about are, I suppose, cases of criminality and, surely, in any nontechnical sense of 'responsible,' people we think of as criminal are, by that very fact, people we *do* think of as responsible for their acts and choices. I can imagine someone being attracted by an argument that goes: anyone criminal is *ipso facto* mad, and anyone mad is *ipso facto* incompetent. But surely the attraction cannot survive reflection on some cases. People commit antisocial acts for all sorts of reasons other than pathology, and if it is hard to draw the line between a psychopath and a mere crook, it does not begin to follow that there is no such distinction or even that there is no such line to draw. Similarly, psychosurgery might be indicated in cases of extreme neurotic anxiety, and it might be admitted that there are *some* decisions which a neurotically anxious patient is incompetent to make. It hardly follows that such a patient is incompetent to make *any* decision; *a fortiori*, it does not follow that he is incompetent to decide whether he wishes to have psychosurgery performed.

Still, there obviously might *be* cases in which psychosurgery is the indicated treatment and where the patient is incompetent to choose, precisely because he is in the state which the psychosurgeon aims to treat. This brings us to the second point under this head which is just that the same sorts of cases arise in other areas of medical practice, so however hard they are to handle, they do not show that psychosurgery raises issues that are simultaneously novel and principled. Unconscious patients, for example, are *ipso facto* not competent to decide, and the fact of their incompetence, of course, raises both moral and legal questions about what may be done for (or to) them. I am not suggesting that it is easy to decide how such issues ought to be dealt with; only that we would have the problem even if psychosurgery had never been invented.

(b) Someone might say that psychosurgery ought *never* to be performed; not even given consent, not even given demand. But I find it hard to see why someone *should* say this, assuming for the moment that he has something other in mind than such issues about side effects as those adverted to above. A worrying case is the following: may someone (e.g., a representative of the state) contract with an otherwise competent agent for the latter to undergo psychosurgery? For example, may the state make such a contract a condition for remission of punishment for criminal behavior? I am assuming that there is no improper coercion on either side, that all the parties to the contract are, in the relevant respects, free agents and, generally, that the usual sorts of conditions on the bindingness of contracts are met. Quite possibly it might be very difficult to determine that such conditions *are* met in the kind of case we are imagining; indeed, it might be *so* difficult as to warrant a sort of legal presumption that they never are. Still, this seems to be only a prudential sort of point, and it is interesting to ask whether there are reasons for supposing such contracts invalid which are not just the sorts of considerations which would invalidate any contract.

The right of contract is, as it were, symmetrical. That is, if it is a limitation on the rights of the state that it may make no contract of the kind we are considering with me, then it is *ipso facto* a limitation on *my* rights that I may make no such contract with the state. But what principle could be appealed to circumscribe my rights that way? Surely I may do what I like with my body. Surely, for example, I could contract with someone to sell them my ear. But if with someone, then why not with the state? And if I can sell my ear to the state, why cannot I exchange it for remission of punishment? And if I can exchange my ear for remission of punishment, why cannot I do, or permit to be done, other things to my body for the same end? (Of course, it would be barbaric for the state to *want* to contract for my ear, but that would seem to be independent of the question whether I have a right to contract for it, assuming that the state *does* want it.)

I am not, I must say, very fond of this argument. But I do not know that my reasons for not liking it have much to do with psychosurgery *per se*. It is not pertinent, for example, that as things now stand, I would be a fool to accept such a contract. (Heaven knows what the actual consequences of psychosurgical treatment might be; and, whatever they were, I would have to live with them.) And it is also not pertinent that one would have reason to doubt the motives of any state that offered such a contract. For, so far as I can see, the right to offer a contract is impugned by one's intentions only insofar as they concern one's intentions to carry out the contract.

If, then, one may not contract for psychosurgery, it must be for reasons which have to do with properties that are specific to psychosurgery; viz., properties that are not shared by other things that one may contract to do, or have done, to one's body. Somewhat analogously, one may not contract to prostitute one's body, and, whatever the reason for this may be, it presumably is something special about *prostitution* that would invalidate such contracts. But what, in the case of psychosurgery, could this special property be? The obvious candidate is that psychosurgery aims to alter one's personality. But it is hard to see how it could be argued that one cannot, in general, contract to do that. Cannot one contract (say with a psychotherapist or, for that matter, with Dale Carnegie) to be made more secure, or more aggressive, or more companionable, or less ill at ease? But if one may make such contracts, then the point about psychosurgery must be that one may not contract to bring about changes in one's personality by *neurological intervention.* But I cannot, for the life of me, imagine what principle such a limitation on one's rights might follow from.

I do not, to summarize, think that psychosurgery is a good idea. For, so far as I can see, our understanding of the mechanisms that it seeks to tamper with is so extremely primitive that only the utmost urgency could justify the tampering. Nor am I at all sure that psychosurgery ought to be legal, for I am dubious that it would be possible to guarantee that consent to psychosurgery is, in practice, both informed and free. But, however the arguments on these points may go, it seems to me that Professor Margolis is right in finding in all of this only 'dialectical,' practical, pragmatic issues. I sense, however, that some parties to this discussion may want to draw a firmer line; that there seems to them to be an issue of principle that I have somehow overlooked. If there is such an issue lying about, I wish that someone would point it out to me.

Massachusetts Institute of Technology,
Cambridge, Massachusetts

NOTE

[1] In fact I do, but this is not the place to exhibit it in detail. Roughly, at least *some* conditions that we think of as psychopathological involve the having of irrational and compulsive beliefs. Now, irrational and compulsive beliefs are likely, by and large, to be false; so our reasons for preferring not to be in such conditions are continuous with our reasons for preferring our beliefs to be true. And that latter preference is not, I think, just a bit of Occidental or bourgeois prejudice. To see that it is not, try to imagine what a mental life would be like if, quite literally and quite generally, one did not care whether any of one's beliefs were true. (These remarks are in lieu of what philosophers call a 'transcendental' argument for the preference for true beliefs. Such an argument would not, I think, be hard to construct. For example, it seems pretty clear that the possibility of understanding the things that people say depends upon assuming that they intend to speak the truth at least most of the time.)

SECTION III

NEURAL INTEGRATION AND THE EMERGENCE OF CONSCIOUSNESS

KARL H. PRIBRAM

MIND, IT DOES MATTER[1]

> Philosophy is akin to poetry, and both of them seek to express that ultimate good sense we term civilization. In each case there is reference to form beyond the direct meanings of words. Poetry allies itself to metre, philosophy to mathematical pattern.
>
> *Alfred North Whitehead*
> *Modes of Thought* ([27], pp. 237-238)

I

The title of this paper, 'Mind, It Does Matter,' is a variant on the old solipsistic saw: "Never mind, no matter." I have always been intrigued by this denial of the mind-brain problem but have found it untenable in pursuing neurobehavioral and neuropsychological research. The results of the research – and I am aware of the criticism that the results of brain research can have no bearing on ontological issues – have led me to a position best described as a biological constructional realism. As a biologist and a physician I can attest to the 'reality' of the psychological as well as the physical constructions that I face daily in laboratory and clinic. My realism is therefore neither naive nor physicalistic. In addition, it differs from critical realism in its emphasis on construction; critical realists are prone to accept their perceptions of the physical world as more or less veridical – the constructionalist is apt to emphasize the relativistic nature of consensual validation.

II. THE PROBLEM

I came to the present stance gradually and via many influences other than research results. Initially, the issue was opened for me when, as an opera-

S. F. Spicker and H. T. Engelhardt, Jr. (eds.), Philosophical Dimensions of the Neuro-Medical Sciences, 97–111.

tional behaviorist, I discussed with Lashley his presention, 'Cerebral Organization and Behavior,' at the Association for Research in Nervous and Mental Disease [14]. It was, it turned out, to be his last published paper. I accused him of having succumbed to mentalism. We argued a bit, and he then accused me of a naive realism. We both immediately realized that we had misjudged the other's arguments – Lashley was not a naive mentalist and I not a naive realist. But neither could we spell out where we differed from the 'naive' position.

I have since read Lashley's remarkable paper many times, and I recommend it as marking the watershed between the behaviorism of the early part of the century and the more encompassing cognitive psychology of today. After detailing some thirty years' effort in clarifying the mind-body problem operationally, Lashley characteristically concludes:

> I was forced to the conclusion that philosophers have been unreliable observers and that much of the difficulty of the mind-body problem is due to their incompetence as psychologists. The phenomena to be explained, as studied by psychologists, are mostly not what the philosophers have claimed them to be...
>
> The problems of mind suggested by the literature fall into two groups: first, the nature of the items, elements or things which are present in consciousness; second, the arrangement or patterns in which these items occur... The problem arising from these two aspects of mind are different. *The content of experience, the sensations and the like, constitutes all that is directly known*. It is the material which has most stubbornly resisted description in the space-time system of the physical sciences. The ordering is not so certainly to be characterized as mental. I shall deal with it first because *interpretation of the content of* experience is derived from the ordering.
>
> *Lashley* ([14], pp. 3-4)
> [Italics mine.]

Note that Lashley is making two apparently incompatible statements: (1) that the content of experience is all that is *directly* known and (2) that the content of experience is *derived* from its ordering. It is this apparent incompatibility which we both sensed and which mystified our discussion. Today, I would challenge Lashley to tell me which statement is, in his opinion, the more fundamental. Since I cannot do this, I can only present my own constructionalist resolution of the issue [21].

III. MENTALISM

But before detailing these observations on the issue of direct perception, some equally fundamental problems need to be faced. One of my original 'missions' in undertaking brain research was to excise mentalism from psychology as my forebears had excised vitalism from biology. The synthesis of organic urea from inorganic carbon, oxygen, nitrogen and hydrogen had sparked the development of biochemistry and the demise of vitalism. I had hoped that the study of brain-behavior relationships, by sparking the development of neuropsychology, would in a similar fashion dispose of mentalism. In a way I feel that my mission is now accomplished but in a completely unexpected fashion. Whereas vitalism proposed some vague 'vital principle' that differentiates the living from non-living, the proposals made by mentalists were of a different order. The richness of mind was not only detailed by professional observers but was experienced by all of us daily. That the phenomena of mind are as real as any other reality is impressed on every psychiatrist who sets out to deal with the social and psychological health of his patients. Mind is *not* vague – mind did not go away when its material 'components' were studied.

To William James is attributed the saying that a 'real' difference is a difference that makes a difference [13]. Mind – the sum of mindings, as Ryle [24] has defined it – matters by organizing the relationship between an organism and its environment. Today, we use the term 'attention' more often than 'minding' for this organizing process – but, mind you, without attention we would not be able to discern what matters.

In essence, I have come to feel that mentalism as an 'ism' will be supplanted by a psychology using definable mental terms of suitable variety. The problem of mentalism in psychology is not at all the same as was the problem of vitalism in biology, and therefore needs a different set of solutions. There is not one vague problem as there was in the case of vitalism; rather, there are specific puzzles such as existential subjective awareness; the projection of sensory events into the 'environment'; the definition of conscious states; the characterization of attention and perception; and of feelings of emotion, motivation and intention, to name just a few. I have also become convinced that behavioral specification of these mental states and processes is not only possible but critical. At the same time, however, such specification leaves unanswered the mind-brain issue with which this conference and my own work is primarily concerned.

IV. REDUCTIONISM, EMERGENTISM AND BEHAVIORISM

For me, the key to the problem lies in the fact that behavior ordinarily describes some organism-environment relationship. Specifications derived from observations of behavior thus assume the skin as a more or less arbitrary boundary. The specifications thus become enriched when the organismic and environmental variables involved can be precisely described. Vision is understood as different from audition when we can properly assign the roles of eyes and ears. Color vision becomes understood scientifically as a mode of perceiving a certain range of the electromagnetic spectrum with retinal receptors containing photosensitive pigments attuned to this range. Vision is a mental term – yet the contributions to the journal, *Vision Research*, would be difficult to fault for their 'mentalism.'

At this point my philosopher colleagues may query: Are not you simply stating a reductionist doctrine? You are confusing an understanding of the pieces with an understanding of the whole. My answer is that this *is* the way of science but – and it is a large *but* – the reductionism is tempered with emergentism. Scientists understand water as a combination of hydrogen and oxygen but stress that the particular organization, H_2O, results in properties which are unique and perhaps even difficult to predict from knowledge of the properties of hydrogen and of oxygen alone. Thus Sperry [26], as many before him (e.g., [12]), see the mind-brain issue in terms of mind as an emergent of brain organization. Our job, therefore, is to determine which brain organizations are responsible for which mental states and processes.

On the whole, I tend to go along with this approach, but as a behavioral psychologist, as well as a brain researcher, I hold that some organizations of psychological states and processes are more readily understood by recourse to their environmental organizers than by delving into the responsible brain organization per se. An analogy helps to clarify what I mean. It is usually easier to understand (i.e., specify the unique characteristics of) a program by studying the organizing 'software' that constitutes the input to a computer than it is by delving into the operations of the hardware switches that compose the working organization of the program per se.

V. INTERVENTIONISM

Reductionism tempered by emergentism and behaviorism thus appears to solve a great many of the mind-brain-behavior problems. But, of course, seemingly new queries immediately come to mind. Are the emergent properties epiphenomena? If so, do we take a dual stance and understand brain organization and the organization of mind to be parallel – in reality – or just parallel ways of talking about identities as becomes a monist? Or perhaps brain and mind interact in the sense that Sherrington [25] suggested: the motor cortex of the brain is where mental organization becomes effective, much as the piano keyboard is where a memorized concerto becomes actualized. The seemingly new queries turn out to be the old ones. The mind-brain issue is still with us even when we understand parts of the problem in terms of reduction, emergence and behavior.

VI. PHYSICALISM AND SUBJECTIVISM

There is yet another issue that is raised by the reductionist approach. Modern quantum and nuclear physics has had to come to terms with the fact that the observations of physical events cannot be excluded from the descriptions of the physical events (e.g., [11], [3]). Thus, if physical description entails observation non-trivially, reduction demands description of the observation and therefore of the observer. The elegant mathematical specifications that make up modern physics are the very mental activities emerging from the physicists' brains (when properly programmed). Physics thus 'reduces' to mathematics which is a mental process. But, as we have already seen, mental processes 'reduce' to brain processes which are physical and which therefore can in turn be reduced to mental...

We can either accept this circularity or go back to our philosophical speculations and laboratory manipulations, ignoring it all as too confusing. I would urge acceptance. Much of what we know comes to us from an ever-widening circularity, especially when the circle is entered from different vantages and with different techniques.

VII. CONSTRUCTIONAL REALISM

I have developed for myself a language that takes account of reduction, emergence, behavior, circularity. Central to that language is the fact that emergence results from particular *organizations* of the reductive constituents. My way of talking about these organizations is to say that, when they become 'realized' or 'embodied,' the emergent properties become manifest.

The concepts 'embodiment' and 'realization' are to be found everywhere in biology. The structure, i.e., the organization, of the genetic potential was found embodied in the DNA molecule. Further realization takes place by way of derepression, RNA, etc., until a viable offspring can be identified.

It may not be so different for an idea. The structure, i.e., the organization, of its potential may be implanted as a memory store in the relations among neurons (an engram). By way of derepression (differential reinforcement), RNA, etc. [18], further realization takes place until a viable manuscript can be identified.

Identification can be at several 'levels.' Thus I can take the offspring and study his behavior ("always yelling, just like his father") or his physical resemblance ("has his mother's eyes") or his brain waves or over the years other realizations of his potential. In like manner, the manuscript can be identified by its immediate effect ("is not this what Lashley meant?"), its physical organization ("a philosophical, not an experimental paper"), its spoken presentation, or its long-term influence.

VIII. EXTRINSIC AND INTRINSIC PROPERTIES OF REALITY

In all instances embodiment – realization – is the result of a process of interaction of one fairly immutable organization with those of a series of varying environments. The immutable organization provides the potential and is *extrinsic* to the medium in which realization takes place. The medium has its own *intrinsic* organization. Realization depends on a highly intricate meshing of these two organizations. I have elsewhere suggested that extrinsic organizations are akin to programs while intrinsic organizations are more likely to be holonomic (hologram-like) in character [19]. The distinction between extrinsic and intrinsic organizations is also similar to the concepts of extrinsic and intrinsic properties as delineated by Russell [23], except that Russell felt that intrinsic properties could never be apprehended. I would rather state that they are difficult – the World War II slogan of the U.S.A. Corps of Engineers is apropos here: "The impossible just takes a little longer." For me it is the intrinsic properties of a recording medium (be it a disc or tape or orchestra) that have to be mastered in laboratory or concert hall that test the patience of artist, engineer, and scientist when they try to realize the extrinsic potential that is the structure of a symphony.

IX. THE SPECIAL CASE OF DIRECT PERCEPTION

Perhaps by now I have either worn thin your patience or have pretty well made clear what I mean by a constructional realism. My realities are, for the most part, painstakingly constructed by me or by others. And you may, on the whole, agree – yet have one reservation. Ideas and thoughts may be constructed as may be physical reality. But what about the percepts which form the building blocks from which all constructions derive?

The issue of direct perception is a currently debated one, in part due to James Gibson's striking experimental demonstrations and refreshing interpretations [7].

Gibson's contribution has been to emphasize that perceptions are not constituted of 'elements' as Lashley (in the quotation at the beginning of this essay) had stated, but come whole and are directly perceived as such. Elements become differentiated as a result of perceptual learning [9]. Gibson further states that the environment 'affords' the organism with the 'information' to be perceived. This 'information' is different for different organisms and depends on the genetic endowment with which organisms scan their environments.

Gibson is not naive in his concept of what constitutes information – this need not be in any simple way isomorphic with the resulting perception. In discussions, however, he is adamant that even if we cannot specify it, the 'information' is present 'out there,' and that if it matches the 'affordances' of the organism, direct perception will occur. Thus Gibson is a realist, and his realism, as well as that of others such as Metzger [15], re-invigorated the position I had originally taken in the discussion with Lashley.

But differences also emerged in the discussion. If 'information' as such can only be perceived through scanning and other activities on the part of the organism (which Gibson calls 'affordances'), and these can change, as when patients with brain lesions are subject to macropsia or micropsia, how then are we to separate 'information' from 'affordance'? (Initially Gibson had no concept of 'affordance' – this was added as a direct result of the discussions – and he thus had a clearer, though, in my opinion, an obviously wrong view of direct perception: "This theory of vision asserts that perception is direct and is not mediated by retinal images transmitted to the brain" ([8], p. 226). I believe that organisms do separate 'information' from 'affordance' by consensual validation, by the mechanisms of projection (see [1], [19]) and by knowledge acquired through learning [10]. These are all constructional processes. In fact, we call a direct perception which does not fit what we know an 'illusion.' Many of Gibson's experimental effects are illusions, as in the centering of the origin of music between two stereophonic sources. Only when direct perception agrees with all we know can we infer that we have perceived the 'information' as it occurs. There is no question in my mind that the 'information' giving rise to a stereophonic effect can be specified, that there exists such 'information' and that the perception is due to that 'information.' I will argue, however, that what is *directly* perceived is on occasion at odds with that 'information' and that laborious search is often necessary to find out why. Direct veridical perception thus becomes a special case where apparent directness matches all we can learn to know about the 'information.'

Further, I claim that eye and brain are necessary to perception. Gibson in the quotation above does not deny this; he merely states that retinal *images* are not necessarily transmitted to the brain to constitute a perception. He prefers to think of the senses as scanners of their input – of the 'information' presented to them. I would argue that scanning is already a constructive operation – and whether one calls the result of such an operation an image (as on a television screen) or not (as in telephone communication) is irrelevant to the discussion.

The issues appear to be these. Gibson abhors the concept 'image.' As already noted, he emphasizes the 'information' which the environment 'affords' the organism. As an ecological theorist, however, Gibson recognizes the importance of the organism in determining what is afforded. He details especially the role of movement and the temporal organization of the organism-environment relationship which results. Still, that organization does *not* consist of the construction of percepts from their elements; rather the process is one of responding to the invariances in that relationship. Thus perceptual learning involves progressive differentiation of such invariances, not the association of sensory elements.

The problem for me has been that I agree with all of the positive contributions to conceptualization which Gibson has made, yet find myself in disagreement with his negative views (such as that on 'images') and his ultimate philosophical position. If indeed the organism plays such a major role in the theory of ecological perception, does not this entail a constructional position? Gibson's answer is no, but perhaps this is due to the fact that he (in company with so many other psychologists) is basically uninterested in what goes on inside the organism.

What then does go on in the perceptual systems that is relevant to this argument? I believe that to answer this question we need to analyze what is ordinarily meant by 'image.' Different disciplines have very different definitions of this term.

The situation is similar to that which obtained in neurology for almost a century with regard to the representation we call 'motor.' In that instance the issue was stated in terms of whether the representation in the motor cortex was punctile or whether, in fact, movements were represented. A great number of experiments were done. Many of them using anatomical and discrete electrical stimulation techniques showed an exquisitely detailed anatomical mapping between cortical points and muscles and even parts of muscles [4]. The well-known homunculus issued from such studies on man [16].

But other more physiologically oriented experiments provided different results. In these it was shown that the same electrical stimulation at the same cortical locus would produce *different* movements, depending on such other factors as position of the limb, the density of stimulation, the state of the organism (e.g., his respiratory rate, etc.). For the most part, one could conceptualize the results as showing that the cortical representation consisted of movements centered on one or another joint (e.g., [17]). The controversy was thus engaged – proponents of punctate muscle representation *vis-à-vis* the proponents of the representation of movement.

I decided to repeat some of the classical experiments in order to see for myself which view to espouse (reviewed in [19], chs. 12 and 13). Among the experiments performed was one in which the motor cortex was removed (unilaterally and bilaterally) in monkeys who had been trained to open a rather complex latch box to obtain a peanut reward [22]. My results in this experiment were, as in all others, the replication of the findings of my predecessors. The latch box was opened, but with considerable clumsiness, thus prolonging the time taken some two- to three-fold.

But the interesting part of the study consisted in taking cinematographic pictures of the monkeys' hands while performing the latch-box task and in their daily movements about the cage. Showing these films in slow motion we were able to establish to our satisfaction that no movement or even sequence of movements was specifically impaired by the motor cortex resections! The deficit appeared to be *task* specific, not muscle or movement specific.

My conclusion was, therefore, that depending on the *level of analysis*, one could speak of the motor representation in the cortex in three ways. Anatomically, the representation was punctate and of *muscles*. Physiologically, the representation consisted of mapping the muscle representation into *movements*, most likely around joints as anchor points. But behavioral analysis showed that these views of the representation were incomplete. No muscles were paralyzed, no movements precluded by total resection of the representation. *Action*, defined as the environmental consequence of movements, was what suffered when the motor cortex was removed.

The realization that acts, not just movements or muscles, were represented in the motor systems of the brain accounted for the persistent puzzle of motor equivalences. We all know that we can, though perhaps clumsily, write with our left hands, our teeth, or, if necessary, our toes. These muscle systems may never have been exercised to perform such tasks, yet immediately and without practice can accomplish at least the rudiment required. In a similar fashion, birds will build nests from a variety of materials, and the resulting structure is always a habitable facsimile of a nest.

The problem immediately arose, of course, as to the precise nature of a representation of an act. Obviously, there is no 'image' of an action to be found in the brain if by 'image' one means specific words or the recognizable configuration of nests. Yet some sort of representation appears to be engaged that allows the generation of words and nests – an image of what is to be achieved, as it were.

The precise composition of images-of-achievement remained a puzzle for many years. The resolution of the problem came from experiments by Bernstein [2] who made cinematographic records of people hammering nails and performing similar more or less repetitive acts. The films were taken against black backgrounds with the subjects dressed in black leotards. Only joints were made visible by placing white dots over them.

The resulting record was a continuous wave form. Bernstein performed a Fourier analysis on these wave forms and was invariably able to predict within a few centimeters the amplitude of the next in the series of movements.

The suggestion from Bernstein's analysis is that a Fourier analysis of the invariant components of motor patterns (and their change over time) is computable and that an image-of-achievement may consist of such computation. Electrophysiological data from unit recordings obtained from the motor cortex have provided preliminary evidence that, in fact, such computations are performed [5, 6].

By 'motor image,' therefore, we mean a punctate muscle-brain connectivity that is mapped into movements over joints in order to process environmental invariants generated by or resulting from those movements. This three-level definition of the motor representation can be helpful in resolving the problems that have become associated with the term 'image' in perceptual systems.

In vision, audition and somesthesia (and perhaps to some extent in the chemical senses as well) there is a punctate connectivity between receptor surface and cortical representation. This anatomical relationship serves as an *array* over which sensory signals are relayed. At a physiological level of analysis, however, a mapping of the punctate elements of the array into functions occurs. This is accomplished in part by convergences and divergences of pathways but even more powerfully by networks of lateral interconnectivities, most of which operate by way of slow graded dendritic potentials rather than by nerve impulses propagated in long axons. Thus in the retina, for instance, no nerve impulses can be recorded from receptors, bipolar or horizontal cells. It is only in the ganglion cell layer, the last stage of retinal processing, that nerve impulses are generated to be conducted in the optic nerve to the brain (reviewed in [19], chs. 1, 6, and 8). These lateral networks of neurons, operating by means of slow graded potentials, thus map the punctate receptor-brain connectivities into functional *ambiences*.

By analogy to the motor system, this characterization of the perceptual process is incomplete. Behavioral analysis discerns perceptual constancies just as it had to account for motor equivalences. In short, *invariances* are processed over time, and these invariances constitute the behaviorally derived aspects of the representation (e.g., [20]). Ordinarily, an organism's representational processes are called *images*, and there is no good reason not to use this term. But it must be clearly kept in mind that the perceptual image, just as the motor image, is more akin to a computation than to a photograph.

We have elsewhere presented the evidence that, for the visual system at least, this computation (just as in the motor system) is most readily accomplished in the Fourier or some similar domain. The evidence is that pattern perception depends, in part, on the processing of spatial frequencies. It is, after all, this evidence more than any other that has suggested the holonomic hypothesis of perception [19, 21].

The perceptual image, so defined, is therefore a representation, a mechanism based on a punctate connectivity that composes an *array*, which is operated upon by other lateral interconnections that provide *ambiences*, which in turn process the *invariances* in the organism's input. What remains to be understood is how such processing gives rise to our perceptions of an objective world separated from the receptor surfaces which interface the organism with his environment.

Von Bekesy [1] has performed a large series of experiments on both auditory and somatosensory perceptions to clarify the conditions that produce projection and other perceptual effects. For example, he has shown that a series of vibrators placed on the forearm will produce a point perception when the phases of vibrations are appropriately adjusted. When such vibrators are applied to both forearms and the subject wears them for a while, the point perception suddenly leaps into the space between the arms.

Other evidence for projection comes from the clinic. An amputated leg can still be perceived as a phantom for years after it has been severed and pickled in a pathologist's jar. A more ordinary experience comes daily to artisans and surgeons who 'feel' the environment at the ends of their instruments and tools.

These observations suggest that direct perception is a special case of a more universal experience. When what we perceive is validated through other senses or other knowledge (accumulated over time in a variety of ways, e.g., through linguistic communication) (see [10]), we claim that perception to be veridical. When validation is lacking or incomplete, we tend to call the perception an illusion and pursue a search for what physical events may be responsible for the illusion. Gibson and his followers are correct, perception is direct. They are wrong if and when they think that this means that a constructional brain process is ruled out or that the percept invariably and directly gives evidence of the physical organization that gives rise to perception.

As noted, there is altogether too much evidence in support of a brain constructional theory of perception. The holonomic model, because of its inclusion of parallel processing and wave interference characteristics, readily handles the data of projection and illusion that make up the evidence for direct perception. The holonomic model also accounts for the 'directness' of the perception: holographic images are not located at the holographic plane but in front or beyond it, away from the constructional apparatus and more into the apparently 'real,' consensually validatable external world.

I have spent some considerable time on this last point because so much of my own research, as well as that of other neurophysiologists, has over the past 25 years shown that the function of sensory receptors is as much under the control of the brain as it is sensitive to environmental input. It is this meshing of the extrinsic 'mental' with the intrinsic 'physical' which constructs reality. So, mind extrinsic, realized intrinsically, *does* matter.

X. CONCLUSION

In concluding, I repeat mentalism has lost its 'ism' to psychology, not to biology. Mental terms are being specified behaviorally. Some of this specification becomes enriched when brain mechanisms are taken into account. This does not mean that brain is ever irrelevant to mind, any more than computers are ever irrelevant to programs. However, programs can be understood, written, and realized on paper or tape instead of in a computer. Some mental processes – e.g., memory storage and retrieval – are also often realized in this fashion. Others such as feelings and perceptions appear to be more immediately brain-dependent for their realization and have therefore become the province of physiological psychologists. Here the mind-brain relationship becomes engaged in daily operations. It is the results of these endeavors which, I believe, are beginning to resolve the Cartesian dualism into a more manageable pluralism. Thus a variety of embodiments of biological organizations are admitted: some are classed as neural, others as behavioral, and still others as perceptual. There are those who would label the organizations themselves as constituting mind (does DNA have a mind of its own?). There are others who infer mind from the behavior of minding, as I have done here. And there are yet others who reserve for the mental, the subjectively experienced. But by verbal and non-verbal communication we come to know something of each other's subjective existence, much as we come to know the existence of the physical world around us. We accept the 'reality' of each, therefore, on pretty much the same grounds – i.e., constructions from patterns of sensory events. So why make more of the duality than there is? This view neither denies the duality nor makes it the rock upon which the unity of knowledge must necessarily founder. In fact, by precisely (even mathematically) describing the operations involved in each realization, a philosophy of mind – in the sense of Whitehead's definition of philosophy – is already in the making.

Stanford University,
Stanford, California

NOTE

[1] This work was supported by NIMH Grant No. MH12970 and NIMH Career Award No. MH 15214 to the author.

BIBLIOGRAPHY

1. Bekesy von, G.: 1967, *Sensory Inhibition*, Princeton University Press, Princeton.
2. Bernstein, N.: 1967, *The Co-ordination and Regulation of Movements*, Pergamon Press, New York.
3. Bohr, N.: 1966, *Atomic Physics and Human Knowledge*, Vintage Press, New York.
4. Chang, H.T., Ruch, T.C. and Ward, Jr., A.A.: 1947, 'Topographical Representation of Muscles in Motor Cortex in Monkeys', *J. Neurophysiol.* **10**, 39-56.
5. Evarts, E.V.: 1967, 'Representation of Movements and Muscles by Pyramidal Tract Neurons of the Precentral Motor Cortex', in M.D. Yahr and D.P. Purpura (eds.), *Neurophysiological Basis of Normal and Abnormal Motor Activities*, Raven Press, New York, pp. 215-254.
6. Evarts, E.V.: 1968, 'Relation of Pyramidal Tract Activity to Force Exerted during Voluntary Movement', *J. Neurophysiol.* **31**, 14-27.

7. Gibson, J.J.: 1966, *The Senses Considered as Perceptual Systems*, Houghton Mifflin Co., Boston.
8. Gibson, J.J.: 1972, 'A Theory of Direct Visual Perception', in J.R. Royce and W.W. Rozeboom (eds.), *The Psychology of Knowing*, Gordon and Breach, New York, pp. 215-227.
9. Gibson, J.J. and Gibson, E.J.: 1955, 'Perceptual Learning: Differentiation or Enrichment', *Psychol. Review* **62**, 32-41.
10. Gregory, R.L.: 1966, *Eye and Brain*, McGraw-Hill Book Company, New York.
11. Heisenberg, W.: 1959, *Physics and Philosophy*, G. Allen and Unwin, London.
12. Herrick, C.J.: 1956, *The Evolution of Human Nature*, University of Texas Press, Austin, Texas.
13. James, W.: 1931, *Pragmatism – A New Name for Some Old Ways of Thinking*, Longmans, Green and Co., New York.
14. Lashley, K.: 1958, 'Cerebral Organization and Behavior', in *The Brain and Human Behavior* (Proceedings of the Association for Research in Nervous and Mental Disease), The Williams and Wilkins Co., Baltimore, pp. 1-18.
15. Metzger, W.: 1972, 'Critical Remarks to J.J. Gibson's Conception of "Direct" Visual Perception, i.e., of Revived Prephysiological Realism', in J.R. Royce and W.W. Rozeboom (eds.), *The Psychology of Knowing*, Gordon and Breach, New York, pp. 233-236.
16. Penfield, W. and E. Boldrey: 1937, 'Somatic Motor and Sensory Representation in the Cerebral Cortex of Man as Studied by Electrical Stimulation', *Brain* **60**, 389-443.
17. Phillips, C.G.: 1965, 'Changing Concepts of the Precentral Motor Area', in J.C. Eccles (ed.), *Brain and Conscious Experience*, Springer-Verlag, New York, pp. 389-421.
18. Pribram, K.H.: 1966, 'Some Dimensions of Remembering: Steps Toward a Neuropsychological Model of Memory', in J. Gaito (ed.), *Macromolecules and Behavior*, Academic Press, New York, pp. 165-187.
19. Pribram, K.H.: 1971, *Languages of the Brain: Experimental Paradoxes and Principles in Neuropsychology*, Prentice-Hall, Inc., Englewood Cliffs.
20. Pribram, K.H.: 1974, 'The Isocortex', in D.A. Hamburg and H.K.H. Brodie (eds.), *American Handbook of Psychiatry*, Vol. 6, Basic Books, New York.
21. Pribram, K.H.: 1975, 'Holonomy and Structure in the Organization of Perception,' in *Proceedings of the Conference on Images, Perception and Knowledge*, University of Western Ontario, May 1974.
22. Pribram, K.H., Kruger, L. Robinson, R., and Berman, A.J.: 1955-56, 'The Effects of Precentral Lesions on the Behavior of Monkeys', *Yale J. Biol. & Med.* **28**, 428-443.
23. Russell, B.: 1948, *Human Knowledge: Its Scope and Limits*, Simon & Schuster, New York.
24. Ryle, G.: 1949, *The Concept of Mind*, Barnes and Noble, New York.
25. Sherrington, C.: 1946, *Man – On His Nature*, Cambridge University Press, Cambridge.
26. Sperry, R.W.: 1969, 'A Modified Concept of Consciousness', *Psych. Rev.* **76**, 532-636.
27. Whitehead, A.N.: 1958, *Modes of Thought*, 3rd ed., Capricorn Books, New York.

MARJORIE GRENE

MIND AND BRAIN: THE EMBODIED PERSON

I

Since, unfortunately, I am not a neurophysiologist, but only a philosopher, I cannot cope with Dr. Pribram's philosophical terminology. I have a hunch that the positions we take on the mind-body problem are convergent, but except for his piquant suggestion at the close (for which I myself have argued elsewhere) that mind is extrinsic and body intrinsic, I do not understand very much of his argument. 'Critical realism' was always over my head, so I have no idea what it is to refute it; and 'isms' in general are something I try, if I can, to avoid. So let me start, not so much from Pribram's general thesis as from the passage by Karl Lashley that he has quoted, and then try to see in relation to some specific examples what the mind-body or mind-brain problem might be said to amount to and what direction we might take – or have taken – if not to resolve it, at least (as Lashley himself in his 1958 paper was professedly attempting) to lessen its tension.

The difficulty in understanding the mind-brain relation, Lashley asserts, stems from the fact that only the content of consciousness is directly known, while the *interpretation* of that content derives from its ordering. It is the content, he holds further, which seems to be stubbornly 'mental,' while the ordering is more readily accounted for in neurophysiological terms.[1] Lashley has not, as Pribram alleges, contradicted himself in this pair of statements, since, he holds, it is the *interpretation* of the content, not the content itself, that derives from its ordering. But what Lashley *has* done (at least in the passage Pribram quotes; he seems to modify his stand somewhat later in the essay) is to succumb to the same temptation to which, he rightly reminds us, philosophers have long been addicted. That is, he has separated something called 'sensations' or the like from their interpretation and the ordering on which their interpretation depends. Now it is precisely the myth of sensations on the one hand and on the other hand the myth of a single

S. F. Spicker and H. T. Engelhardt, Jr. (eds.), Philosophical Dimensions of the Neuro-Medical Sciences, 113–129.

unitary consciousness which somehow 'contains' them that has led philosophers, psychologists and even some neuro-physiologists into the tangle of arguments and counterarguments that have plagued and, it seems, continue to plague us. True, whatever may befall psychologists or neurologists, no post-Kantian philosopher *ought* to have succumbed to this temptation; but as Kant should have been the first to recognize, what ought to be is not always what is. I want to argue here that once we have rid ourselves of the incubus of sensations, we can consider our various ways of making sense of the things around us (1) not primarily in terms of 'minds,' but of 'persons,' and (2) not in terms of a simple 'unity of consciousness' as the sole criterion of personal identity, but rather in terms of a narrative model of the person which is grounded both in our bodily nature and in our cultural and historical reality.

II

But let me, as I have promised, get down to cases. One of the recent developments in neurophysiology that philosophers have been much taken with concerns cerebral disconnection and its clinical consequences. Let me start with the by now classical case of P.J.K., the patient of Geschwind and Kaplan reported in 1962, and remind you briefly of their observations, as well as of the philosophical problems which, according to Geschwind, these observations suggest ([6], pp. 22-41). Sperry's work with Drs. Bogen and Vogel's commissural patients uses a more sophisticated experimental apparatus, but Geschwind and Kaplan give us a more detailed case history, and it seems to me also that the way Geschwind poses the possible philosophical challenges, in the papers I shall refer to in a moment, is clearer than the concluding section of Sperry's 1970 paper [21]. Moreover, I need repeat only a fragment of the observations described by Geschwind and Kaplan in order to lay the groundwork for my argument. They report:

> With his eyes closed, the patient correctly and rapidly named objects and cardboard letters placed in his right hand. Similarly, he could, after the object was taken away, correctly draw the object which had been in his hand. If letters or numbers were traced on his right hand, he generally named them correctly... By contrast, when objects, numbers, or letters were placed in the left hand, the patient (with eyes closed) generally named them incorrectly. The errors usually bore no resemblance to the stimulus object. For example, a ring was identified as "an eraser," a watch as "a balloon," a padlock as "a book of matches," and a nail as "an elastic." The errors were

not constant. Thus, on another occasion he called a ring "a package of some sort"; a screw driver on one test was "a spoon" and on another "a piece of paper."

It was striking to observe that when handling the objects he would move them actively about within his fingers and would focus on their salient features. For example, he inserted his finger into a thimble, ran his thumb over the teeth of a comb, rubbed the bristles of a tooth brush, and retracted the point of a ballpoint pen. Despite his appropriate handling of the dominant features of these objects, he misnamed them all.

When it was made clear to him that he was to show how the object placed in his hand was to be used, the subject (with eyes closed) proceeded with little hesitation to manipulate the item correctly, usually giving simultaneously an incorrect verbal account. Thus, given a hammer, he made hammering movements correctly while saying "I would use this to comb my hair with it." Given a key, he went through the motions of inserting it into a lock and turning it but said that he was "erasing a blackboard with a chalk eraser." Holding a pair of scissors correctly, he made cutting movements in the air but said that "I'd use that to light a cigarette with." When his eyes were open he correctly identified the objects in his left hand ([6] pp. 28-30).

Further,

If an object or letter was placed in the left hand behind the patient's back, he generally was unable with the right hand (1) to select it correctly from a group of objects presented to him either visually or tactually, (2) to write the name of the object, or (3) to draw a picture of the object. For example, a letter which would have been drawn correctly with the right hand after having been held in the right hand was drawn incorrectly with the right hand if the letter was placed in the left hand. Similarly, if objects were placed in the right hand, the patient was usually incapable of performing any of these maneuvers with his left hand ([6], p. 30).

We need not concern ourselves here with the detailed diagnosis of the patient's pathology, which was correctly predicted by Geschwind and Kaplan and confirmed by post-mortem examination; it suffices to say that there was disconnection of 80% of the corpus callosum. The question that concerns us is the question: What, if any, philosophical reflections does such a case suggest?

Geschwind, in his 'Disconnection' paper of 1965 and again in a 1967 paper on the apraxias, lists three areas where conceptual questions arise, of which the first and third seem to be rather clinical than philosophical ([16]), chs. VIII and XIV). First: the question of the whole patient. This is a practical matter. Of course the patient *was* a whole person, even if his symptoms resulted from callosal disconnection – and I am sure that when Geschwind or Kaplan drove him from home to the hospital for outpatient observation (as they report that one of them did) he (or she) treated him as such. What is in question here is the demand for a holistic approach in clinical practice like that advocated by Kurt Goldstein, though, as

Geschwind has pointed out elsewhere, abandoned even in Goldstein's own interpretation – and treatment – where localized areas of the brain rather than global conditions seemed to be involved ([6], ch. V). Geschwind's third point, the reliability of introspection, is again a practical matter. Of course if the speech areas are isolated or damaged, introspective reports will not be reliable. But the problem that teases the philosopher's imagination is the second one Geschwind raises: the question of the unity of consciousness (one of the problems listed also by Lashley in the paper Pribram cites). Surely the patient is feeling normally, with his left hand. All his behavior indicates that he is doing so and, indeed, special tests were devised to show that this was the case ([6], p. 26). But this perception does not get through to the left brain. Is the patient somehow 'of two minds'?

Some philosophers have drawn on the work of Geschwind, Sperry, Gazzaniga and others to suggest that more abstract judgments may be distinguished from more 'intuitive' judgments, or what we 'spell out' hedged from what we would rather not recognize in our own 'consciousness': that such complexities of our mental lives may be correlated (perhaps even identified) with the asymmetry of cerebral organization (cf., e.g., [5]). These are comforting neurological underpinnings for philosophical arguments, and I am all for them. But my question is a different one. Does the unity of consciousness *have* to be thought through in a new way in the light of cases like that of Geschwind and Kaplan, and if so, why and how? A colleague with whom I was discussing Geschwind's statement insisted that *philosophically* the case of P.J.K.'s discordant behavior, and seemingly discordant experience, is no different from the old Berkeleyan case of the two hands, one pre-heated, one pre-cooled, placed in a basin of tepid water, where one feels hot and the other cold (a case nicely photographed in Richard Gregory's *The Intelligent Eye*; I will have more to say about his interpretation of it in a moment).[2] It is true that the subject in this case has indeed two 'consciousnesses,' in the sense of two 'sensations,' one hot, one cold. What did Berkeley infer from this situation? Not: 'never mind,' but only: 'no matter.' And could not he have said in the Geschwind-Kaplan case also something like this: usually the algebra of ideas God has given us to work with ('ideas' identical with what the vulgar call 'objects') guides us pretty well, but this poor Boston cop was misguided because the objects = ideas that constituted his brain were damaged and so the mind somehow connected with this set of ideas-or-objects contained peculiar ideas-or-objects instead of normal ones? I had to admit this would be a *logically* possible interpretation; but it seems to me, and would to most of us, I hope, a terribly far-fetched one, just as Berkeley's own interpretation of the tepid water experiment appears to me (or us) a terribly far-fetched one. The difference, it seems to me, comes from the difference in what we, in our historical-intellectual situation, take for granted and what seems questionable or difficult to account for. Berkeley 'knew' – thought he knew – there were sensations, as Lashley still affirmed, at least in the passage Pribram quotes; he also 'knew' – thought he knew – mind as substance and as agency: himself as the cause of his actions, a relation which appears to us, I think, a very mysterious and complicated one, not at all self-evident. I, for one, find it very hard to tell when I have really done something and when I have been pushed. Finally, Berkeley 'knew' – thought he knew – the existence of an all powerful, all-good Deity who had arranged all these ideas = objects so that, other things (like sins or brains) being equal, we would be guided adequately through this vale of tears and reap our reward hereafter. But such an all-powerful and all-good Deity is a self-contradiction, as Hume, following Epicurus, amply proved, and most of us, I surmise, would not care to interpret Geschwind and Kaplan's data (or Berkeley's either) with His help. (If Bishop Berkeley *is* at present casting golden crowns upon a glassy sea, I hope, in the hell I will soon be frying or freezing in, I shall at least be vouchsafed that knowledge.) In any case, to Berkeley it was utterly clear that there is God, there are minds which are active agents, there are ideas which they passively contain. It was *not* clear to him that there are material things distinct from our – or any – consciousness. For us, on the other hand, it is clear that there is a natural world of which we, as animals, form part. How our 'minds' are organized, what 'agency' is – let alone what God is, if we are still willing to entertain the question – is what has become problematic. And that there are sensations, isolable, identifiable, single 'bits of consciousness' out of which experience gets built is just plumb *wrong*. Why so many intelligent people in the eighteenth century – even David Hume – believed in them I will never know unless some brilliant intellectual historian tells me. But certainly our everyday experience just *is* not made up of such 'psychological atoms.' Even a case like the two-hands one entails a context; it is precisely the context that gives the appearance of paradox, and the further, temporal context that alleviates it. Even Philonous or his poor stooge Hylas could not have obtained this baffling pair of feelings if he had not first kept one hand in hot water and the other in cold – a mere manipulation of ideas? Nonsense! The hidden metaphor in 'manipulation' sharpens the point. Moreover, the denial of such a psychological atomism is part, at least, of what Geschwind and Kaplan's observation of their patient suggests as well. The human brain, 'normally' functioning, is a pluralistically organized, immensely complex machine used by the human person possessing one for the unitary apprehension of some aspect or other of the world around him, which, other things being equal and in particular the corpus callosum being intact, he can also report in reliable, unitary fashion. Human minds as they mature are the uses made, in cultural contexts, of these pluralistically organized softwear machines. What *I* take to be obvious, not logically necessary but acceptable to any reasonable person, is (1) the existence of an 'external world' which contains me even though it does not directly feel my feelings (a world external to my lived body, not external to some non-fleshly black-box type feeling apparatus 'inside me'), (2) the realistic thrust, in general, of my perceptions and reports of perceptions, even though they are not incorrigible, (3) the uselessness, if not absurdity, of *any* concept of substance, whether mental or material, (4) the contextual – already organized – nature of almost all, if not all, so-called 'immediate' perceptions, (5) a difference of *some* kind between my actions (and those of other intelligent agents) and the motions accounted for by the explanations of physics and chemistry – but a difference that takes some pretty complicated thinking to explain it and is by no means 'self-evident' either in general or in any particular case. (God we can leave out of it: that there is in some sense some reality expressed by or corresponding to the religious experience of religious persons I do not want to deny; but that a theory of perception includes, or should include, reference to some such reality – as it does in Berkeley's, or for that matter Descartes' case – I *would* emphatically deny. Indeed, it is the involvement of Deity that makes their theories of perception so unsuccessful and so quaint.)

What, then, in terms of these very different presuppositions do *we* make of Berkeley's case of the two hands in tepid water? Gregory's interpretation is as follows:

One hand is placed in a bowl of cold water, the other in hot. If the hands are then placed together in a bowl of tepid water the water feels, and is judged to be, at the same time warm for one hand and cool for the other. The cold and hot water have selectively adapted the hands, upsetting their calibration, so that temperature is then judged incorrectly by each of them. Although we know intellectually that water cannot be both hot and cold at the same time, this is how we feel and judge it to be. The brain does not reject the paradox.

We experience something we know to be physically impossible ([9], p. 74). This statement is disquieting. On the one hand, 'we feel and judge,' on the other, 'we know intellectually.' But judgments are judgments. Is a perceptual 'judgment' radically different from an 'intellectual one'? I find myself here in a dilemma – or perhaps just in a muddle. For on the one hand I want to accept Gregory's general thesis that 'perceptions are hypotheses,' but not his acceptance of Helmholtz's view of them as 'unconscious inferences.' There just is not any 'inference' in the hot and cold hand perceptions. We feel them that way and report them that way, if asked to do so, just as we see the lines in the Müller-Lyer illustration as unequal and report this perception, even though the ruler shows us they are equal. 'Illusory perceptions' are mistaken hypotheses that we cannot help entertaining alongside (once we have other evidence) the hypotheses or metahypotheses that they are incorrect. True, they are hypotheses forced on us by circumstance and the nature of our nervous systems, and in this distinguished from the more loosely held hypotheses which we can correct either through perceptual or conceptual learning. But I do not see where, in the perceptual case, the 'inference' comes in. All the way from perception to the most refined theoretical knowledge or artistic insight we are trying, out of and within our given biological-historical situation, to make sense of some aspect of experience, whether closer to sensed experience or more theory-laden and further abstracted from it. Some of these 'appraisals,' as Polanyi calls them ([18], passim, see index), or instances of 'sizing-up,' to use Wallace Matson's term [19], have been railroaded by evolution into fairly rigid, and occasionally deceptive, paths – which must, however, on the whole have been reliable guides to coping with the real world, else natural selection would have eliminated them – or us. Others, thanks to the loose and complex organization of our brains with their non-limbic associations most highly characteristic of our species, and thanks to our development, within one culture or another, as initiators and users of highly structured symbol systems, are more flexibly connected to sensory input and more easily susceptible of learning, criticism and correction. Only in cases like that of P.J.K. do we notice, through the oddities consequent on callosal disconnection, how delicate and perilous is the kind of 'intellectual' coping we usually do – and even it, of course, is not only brain-mediated but percept-mediated as well. Perceptions are hypotheses, yes. Yet, contrariwise, all hypotheses are grounded in perception, from which they spring and to which they must ever again return. Perception as hypothesis, or perception as 'information' actively selected, as Gibson, in a position at least partly convergent with Gregory's, interprets it, is both the paradigm and the ground of all our knowledge [7. 8]. *Perception*, mind you: full, *'sinnvoll,'* world-oriented perception, not those mythical 'givens,' in the demolition of which philosophers have wasted almost as much time, breath and ink as they did in their defense. But let me not waste more.

One more point, though, in interpretation of the case of P.J.K. and the question of the 'unity of consciousness.' Pribram (to add one more '-ist' if not '-ism' to his repertoire) has elsewhere referred to his 'solution' of the mind-brain (or mind-body) problem as the 'biologist' solution [19]. If this means that mental existence is embodied, so far, so good. But, in the case of human persons at least, mental existence can be achieved, not through neurological forces and processes alone, or even bodily forces and processes in general, but only through the complex interplay of the person's bodily being and the artifacts: family structure, social and political institutions, languages, art forms, rituals, that have both permitted his development to take the shape it has taken and prevented it from taking any of an indefinite number of alternative, but no less human, forms. Look again at Geschwind and Kaplan's patient. True, a medical history is only a very partial and truncated 'history,' but even it tells something: "P.J.K., a 41-year old married white male Boston police officer" ([6], p. 23). Forty, not 20 or 60; married, not single; white, not black, Oriental, Mexican-American or Puerto-Rican; male, not female; Bostonian, not Parisian, for example; a member of the police force, not (presumably) a pusher or a tap dancer or a neuro-surgeon. Maybe indeed even this brief description informs us of more than it says. I have been told that I came home from my first day at kindergarten in New Haven and asked, "Mother, are we Catholics or Republicans?" The person lives somehow at the center of what the complex organization of his body permits and the complex organization of his social world (or worlds) demands. This confluence of shapes and forces any simple identification of 'mind' with specifiable bits of consciousness at one extreme or with a simple unity of consciousness at the other extreme necessarily fails to capture. And if a subjective unity of feeling or of feelings fails to do the job, so much the less will the concept of an objective physical, even biological, unity do it, at least not on its own. Only the historical span of this lived body-in-its-world can identify a person as this person and no other. The loosely but subtly interwoven structures of the human brain provide one set of necessary conditions for such an existence. The more or less loosely but subtly interwoven structures of the human world provide another set, equally necessary. 'Mind,' if you want to call it so, develops in the tension between the two. But this is not the opposition Lashley suggests: between a 'content of consciousness immediately known' and an 'interpretation' of it derived from its neurophysiological ordering. The content is already interpreted – as we can tell when, as in P.J.K.'s case, its 'normal' ordering breaks down. In Kantian terms, experience is always already a synthesis of the manifold. Moreover, the content as lived, is lived bodily (an insight alien to Kant's still Cartesian way of thinking). And on the other side, the ordering has been from the beginning not only electrochemical, but at the same time artifactual and symbolic. It is the unity of being-in-a-world, always ambiguously psycho-physical, always ambiguously public-private, always ambiguously active-passive, that provides the ground of personal identity. And even when, as in both our cases, a simple 'unity of consciousness' breaks down, routinely in sensory illusions like the tepid water one, or more dramatically and anomalously in the perceptions and reports of the callosal patient, there is recognizably one person whose history includes these less or more puzzling features. Like Hylas or Philonous, P.J.K. continues to cope, up to a point, with the situations of ordinary life – and Bogen and Vogel's patients, relieved of severe epileptic seizures, presumably coped, on the whole, better after than before surgery. In other words, we acknowledge in each case the presence of a person, the anomalies of whose perceptual experience point up the subtlety with which our 'normal' dealings with things in our world are organized. Here is a personal history with some kinks to it, sufficient sometimes to disturb the patient himself, as reported, for example, in Goldstein's case of 1908 ([6], ch. V, pp. 62-72), or in the case of Vogel's first commissural patient, a very mild-mannered man, whose left hand, after surgery, occasionally reached out and slapped his wife, to his own surprise and embarrassment. Yet this is still a personal history, even if an odd one, not no personal history at all.[3]

III

So far I have responded both to the contrast posed by Lashley and to the question raised by Geschwind by sketching what I take to be some reasonable assumptions about personal identity in its relation to traditional views of immediate experience and of a consciousness which is said to unify it. One can hardly question the 'unity of consciousness,' in our day and time, however, without acknowledging the effect on this venerable concept of the impact of Freud and the notion of something – or some process or set of processes – called '*un*conscious.' The person is not to be, in any simple way, identified with a unified consciousness, not only because of the pluralistic organization of his brain, not only because of the pluralistic organization of his social world, but because much of his mental life – much of what makes him a person – is either preconscious or unconscious: in the latter case accessible to consciousness either not at all or only through the esoteric techniques of psychoanalysis. But if, as a student of philosophy, I am ignorant of neurology, I must confess to an even more profound ignorance of psychoanalytic literature and practice. In a mood like that of Kant at the beginning of the Dialectic, therefore, I hesitate to leave the firm ground of cerebral connections and disconnections for the vast and, to me, uncharted sea of depth psychology. Quickly, however, I must dive in and hope to surface again before suggesting, in conclusion, a direction for philosophical reflection in the face of the questions I have raised so far. Let me take my quick dip at one special (and I hope not too dangerous) place: suggested both by Robert Solomon's recent essay in the Anchor Freud volume [20] and by the first part of Alasdair MacIntyre's argument in his book *The Unconscious* (published, coincidentally, the same year as Lashley's paper [13]). Both Solomon and MacIntyre stress (as Pribram has also done in a paper Solomon refers to) the importance of Freud's early neurologically oriented *Project*. Freud himself disowned, and even tried to burn, this early MS, not, however, it seems, because he wanted to disown a neurophysiological foundation for his psychic apparatus of id, ego, superego, but only because the neuroanatomy of the time was inadequate to support it. What matters for my present thesis in this piece of history is that Freud's rejection of the traditional identification of mind with conscious mind did not signal a return to classical dualism with the psychic system of energy, repression, sublimation and what not as a free-floating set of non-physical events or entities. Rather, as Solomon stresses, Freud's use of quasi-metaphorical, quasi-physical terms suggests an awareness that it is forces in the human organism, and therefore materially existing, ultimately neuro-physiological processes, that develop, or are frustrated and diverted in their development, in the patterns Freud describes, or something like these – I am not suggesting that we swallow all his mythology whole. But of course, on the other hand, it is clear on the face of it that the 'forces' Freud is describing are also psychological and social forces, impacts of the human world upon and assimilations of the human world in the advancing, or in the neurotically diverted or psychotically shattered, lived experience of the individual human organism. Freudian theory stands, therefore, it appears, between the lived, neurologically founded experience of the person (and, indeed, of the species) and the lived structures of the human world which he strives to integrate successfully or unsuccessfully and so produces his own individual bodily-symbolic, psychical-embodied history. Again, this entails neither the rejection of embodiment nor the rejection of mental existence as the organized, and organizing, interpretation of experience but acknowledgment of the person as a task: placed irrevocably between the two inextricable poles of his existence. It means reading 'mind' as the uses of the body as mediator of and participant in the structures of psychical and interpersonal reality.

IV

Where, finally, as philosophers, do we go from here? A recent paper by Tom Nagel, "What is it Like to be a Bat?" [16], begins: "Consciousness is what makes the mind-body problem really intractable." I hope he meant this to be funny. Clearly, were it not for consciousness we would have no problems at all, and were it not for self-consciousness, or reflective consciousness, a consciousness of being conscious, we would have no philosophical problems, including that of mind and brain or mind and body. But, at the same time it is not funny: it is on the one hand the concept of immediately known units of consciousness that Lashley mentions and on the other the concept of a simple and straightforward unity of consciousness of the kind questioned by Geschwind that has misled philosophers (as well as some neurologists) and has complicated the task of transcending the kind of over-simple, hypostatizing mind-body dichotomy bequeathed to us by Descartes. Either (as Heidegger argues in 'The End of Philosophy') philosophers have enclosed themselves in a cage of subjectivity from which they can find no escape to the realities of the human, let alone the natural, world [12]. Or, (chiefly in the British tradition, which Heidegger ignores) while they have indeed sought the contrary of the subjective, that is, the 'scientific' or 'exact,' they have found, paradoxically, the 'objective' shrinking to sensations – as happened to Hume, for example, and in his wake to Russell. So the seekers for objectivism either become more subjective than the subjective philosophers themselves, or, in despair (like Quine or again like Russell) they throw the subjective, and with it consciousness, overboard altogether and cling dogmatically to the laws of physics or behavioristic psychology (how known by whom?) as their only support. Nagel's cure for this last and indeed deplorable move is, he says, to reinstate the subjective. There must be something it *feels* like to be a bat and something else it *feels* like to be human. But it's not just a question of *feeling*: that is the whole point. It is a question, to put it in quasi-Aristotelian terms, of what it is for a bat to be as distinct from what it is for a human being to be. To revert to feeling, or the subjective, or to consciousness as the medium of our inquiry would be to fall back into the very traps set for us by the past three hundred years of philosophical thought.

There is no need for such a relapse, however. In my discussion so far I have been drawing, without naming them, on a number of contemporary sources for a different approach to, or better, if you like, evasion of, the traditional problem. I have suggested that we take 'person' rather than 'mind' as our central concept. This move is familiar to Anglo-American philosophers chiefly through the work of P.F. Strawson [22, 23]. Strawson identified persons as individuals to whom both 'material' and 'person' predicates apply. This was a radical move in the empiricist tradition: but still, I think, inadequate to take us safely beyond the perilous shoals in which Hume grounded us or the equally perilous depths of more Teutonic speculation. What one needs further, as a necessary condition for an adequate philosophy of the person, is a distinction between the subjective and the personal, a distinction which is carefully drawn in Michael Polanyi's *Personal Knowledge*, but so presented, I am afraid that most of his readers have failed to understand it ([18], passim, pp. 300-303, 324, 326). The subjective, in Polanyi's usage, is that minimal aspect of experience which can only be authenticated, but neither verified (as in empirical statements making objective claims), nor validated (as in mathematics or the arts). If a neurologist tests my skin for normal sensation, only I can authenticate for each pin prick the claim "I feel it" or "I don't feel it." This is the kind of 'subjectivity' Lashley and, I take it, Nagel are referring to. It constitutes (except in moments of intense pain or ecstasy) a very minor strand of experience, relatively 'inward,' relatively passive, extremely evanescent. The personal, in contrast, is the whole, organized span and spread of an individual history which I have already referred to. It is at once active and passive, at once mental and bodily, at once public and private. Further, that the being of the person is being-in-a-world is a thesis drawn in the first instance from Heidegger's *Being and Time* [11]; but the emphasis on bodily being (*lived* bodily being) on the one hand and immersion on the other hand in a *common* human world is drawn rather from Merleau-Ponty's *Phenomenology of Perception* than from Heidegger's non-sexed and when authentic (I believe) non-social *Dasein* [15]. Many of my locutions, further, derive, or could be derived, from Helmuth Plessner's philosophical anthropology [10, 17]. That one needs to interpret the person's being-in-the-world in an historical or narrative mode is a thesis, again, supported by philosophical argument from a number of sources. In a way, it is Heideggerian, although here again I find Merleau-Ponty's approach less strangely askew than Heidegger's concept of destiny. Moreover, the notion of something like a 'narrative model of man' I have heard proposed independently by A.C. MacIntyre in Boston and G.H. von Wright in Jvaskyla in Finland.[4] Clearly there is a road to travel, and partly already well-travelled, that leads to the alleviation of the tension described by Lashley or the puzzle posed by Geschwind.

In conclusion, however, I want to make what is perhaps a more heterodox, but yet, I hope, pertinent proposal. Not at variance with the sources already mentioned, but in support of them, I want to suggest that, relative to the notion of either units of consciousness (in the tradition of empirical philosophy) or of a unity of consciousness (in the traditional concept of the 'self' as a kind of entity), what is needed – and what has already been happening in some of the literature I have all too sketchily mentioned – is what Jacques Derrida calls a 'deconstruction' of the tradition (e.g., [1], [2], [3]). True, Derrida is dealing, for the most part at least, with the more speculative, chiefly Continental strain of philosophizing, while I have been speaking here largely in terms consonant with the empirical branch of our divided Cartesian heritage. His proposal, however, should apply, as I hope to show you, to both styles of argumentation.

What does Derrida mean by 'deconstruction'? To tell you even very briefly I shall use language that will probably sound very strange to you. Traditionally, Derrida believes, philosophers have longed for, and even alleged that they have found, a presence of the mind to the really real, a univocality and luminousness in living thought and in its spoken expression, which can bring our hope of ultimate understanding to fulfillment. This hope of Presence, he insists, has persisted throughout the history of Western thought, all the way from Plato's claim to know the Forms, almost three thousand years ago, to the thesis underlying modern linguistic theory and phenomenology. What we need now is to 'deconstruct' this claim: not to demolish it – that's hopeless – but at least to pry ourselves loose from it, to acknowledge a different aspect of language and of meaning. What we need to stress is not the alleged Presence of the mind to reality but, on the contrary, the gap between meaning and meant, saying and said, that characterizes all discourse and all thought. Writing, it has been claimed from Plato onward, is a bad second best to speech: it is dead speech or, as Saussure put it, 'language in drag' ([1], p. 52). But, in Derrida's view, the externality of sign to signified that is symptomatic of writing in fact characterizes all language, spoken as well as written, and indeed experience itself, which, after all, is organized, as the structure of the brain suggests, in traces of traces, never at one with itself in the fashion that the dream of Presence demands. His central theme, therefore, is what he calls 'difference,' the temporalizing, and temporizing, of experience, its differing from itself in the sense of contrariety or conflict, the gappiness within and beyond the synthesis that, on the surface, gives it meaning and form ([4]), pp. 1-29). We cannot escape the hope of Presence, as we cannot escape, in some form, a demand for a unity of consciousness as what unifies a human life. But we must complexify that notion, so to speak, allow it its roots both in lived reality and in artifact, stretch it to meet the conflicts, pluralities, byways that confront us equally when we consider (as, admittedly, Derrida is usually far from doing) the intricacies of cerebral organization or the complexities of the human world within which a given human brain in its given human body achieves its particular history and to which, in its unique way, it gives expression.

Derrida, as I have indicated, is trying to pry philosophical thinking loose from what is basically a Platonic attitude: the hope of the soul's (or the mind's) vision of the Forms. More 'down to earth' reflection, in the Anglo-American vein, however, stands just as much in need of a parallel, 'deconstructing' move. One may plausibly argue, indeed, that that is what the later Wittgenstein was trying to do for it; that is another story. But let me return here, finally, to another passage in Gregory's *The Intelligent Eye* which suggests (a contrario) the direction in which, I believe, we need to move. Granted, Gregory is a psychologist, not a philosopher, but the position he takes is only too typical of much philosophical doctrine, and I venture to take it as such. He is speaking, in the passage I want to refer to, of what appear to him the unique achievements of man or, at least, of modern 'scientific' man. There are two such achievements, he tells us. The second is measurement, which seems relatively trivial and, as a major force, local to modern Western science. The other is language, which allows, in Gregory's words, 'logical and numerical relations to be expressed without ambiguity" ([9], p. 150). It permits, in other words, a precision otherwise impossible. But such a thesis represents the traditional 'Platonic' delusion from the other side or, strictly, the Aristotelian or Cartesian delusion, for Plato knew only too well the necessary imprecision of language. Is it really literal, exact reference that language facilitates?

Let us return for a moment to Geschwind and Kaplan's patient. If (with eyes closed) he held a key in his right hand, he could name it correctly, whereas if he held it in his left hand, he would say at random, "It is a rubber band" or "a comb" or what you will. Yet if he drew the key with his right hand, corresponding to the left, speech-endowed hemisphere, he would make only a rough outline, while if he had held the key in his left hand, although he could not name the object correctly, he would draw it with extreme precision. The left brain's achievement of language, this suggests, functions not to produce precision, which belongs rather to the more 'literal-minded' but inarticulate right brain, but rather to permit a kind of 'tricky imprecision' (to borrow a phrase of Geschwind's).[5] Or, to quote Geschwind again, language works by producing data reduction at the expense of precision. Imprecision is of its essence. In Derridian terms, speech is a power of producing, not so much exact or literal meaning, as metaphor ([4], pp. 1-29). It is a power of drawing together the unlike which are yet somehow or other the like. It is a rather diffuse and messy capacity well suited to coping in a rather diffuse and messy way with a rather diffuse and messy world. Philosophy, under the spell of science or of an over-simplified model of science, needs deconstruction just as urgently as does philosophy under the spell of traditional metaphysics, perhaps more urgently, because it speaks, in our day, with more authority.

University of California,
Davis, California

NOTES

[1] Lashley, K.: 1958, 'Cerebral Organization and Behavior', *The Brain and Human Behavior; Proceedings of the Association for Research in Nervous and Mental Disease*, The Williams and Wilkins Co., Baltimore, pp. 1-18, quoted by Dr. Pribram in the preceding chapter.

[2] Berkeley, *Three Dialogues Between Hylas and Philonous,* first dialogue. The comparison was suggested to me by Professor Michael Wedin of the Department of Philosophy at the University of California at Davis. I am most grateful to him, as well as to Professor Fred R. Berger of the same department, for discussions of the problems dealt with in the present paper.

[3] Geschwind, N., personal communication. Dr. Geschwind was commenting on an interview with this patient.

[4] Lectures delivered by A.C. MacIntyre at Boston University in the fall semester, 1972; and lecture by G.H. von Wright at the University of Jvaskyla in the summer of 1973.

[5] Geschwind, N., personal communication. Dr. Geschwind reported this incident, which was not included in the published paper. I should like to express my gratitude to him for reading an earlier draft of the MS. and discussing it with me.

BIBLIOGRAPHY

1. Derrida, J.: 1967, *De La Grammatologie*, Éditions de Minuit, Paris.
2. Derrida, J.: 1972, *La Dissémination*, Éditions du Seuil, Paris.
3. Derrida, J.: 1967, *L'Écriture et la Différence*, Éditions du Seuil, Paris.
4. Derrida, J.: 1972, *Marges de la Philosophie*, Éditions de Minuit, Paris; trans. by D. Allison in 1973, *Speech and Phenomena*, Northwestern University Press, Evanston, pp. 129-160.

5. Fingarette, H.: 1969, *Self-deception*, Routledge and Kegan Paul, London; Humanities Press, New York.
6. Geschwind, N.: 1974, *Selected Papers on Language and the Brain; Boston Studies in the Philosophy of Science*, vol. XVI, Reidel, Dordrecht.
7. Gibson, J.J.: 1966, *The Senses Considered as Perceptual Systems*, Houghton Mifflin, Boston.
8. Gibson, J.J.: 1972, 'A Theory of Direct Visual Perception', in J.R. Royce and W.W. Rozeboom (eds.), *The Psychology of Knowing*, Gordon and Breach, New York, pp. 215-227.
9. Gregory, R.: 1971, *The Intelligent Eye*, McGraw-Hill, New York.
10. Grene, M.: 1974, *The Understanding of Nature; Boston Studies in the Philosophy of Science*, vol. XXIII, Reidel, Dordrecht, chs. XVIII and XIX.
11. Heidegger, M.: 1927, *Sein und Zeit*, Niemeyer, Tübingen.
12. Heidegger, M.: 1969, 'Das Ende der Philosophie und die Aufgabe des Denkens', *Zur Sache des Denkens*, Niemeyer, Tübingen, pp. 61-80.
13. MacIntyre, A.C.: 1958, *The Unconscious*, Routledge and Kegan Paul, London.
14. Matson, W.: 1976, *Sentience*, University of California Press, Berkeley and Los Angeles.
15. Merleau-Ponty, M.: 1945, *La Phénoménologie de la Perception*, Gallimard, Paris.
16. Nagel, T.: 1974, 'What is it Like to be a Bat?', in *Phil. Rev.* 83, 435-450.
17. Plessner, H.: 1970, *Philosophische Anthropologie*, S. Fischer, Frankfurt, with 'Nachwort' by Günter Dux.
18. Polanyi, M.: 1958, *Personal Knowledge*, University of Chicago Press, Chicago.
19. Pribram, K.: 1971, *Languages of the Brain*, Prentice Hall, New York.
20. Solomon, R.C.: 1974, 'Freud's Neurological Theory of Mind', in R. Wollheim (ed.), *Freud: A Collection of Critical Essays*, Doubleday, New York, pp. 54-72.
21. Sperry, R.W.: 1970, 'Perception in the Absence of the Neocortical Commissures', *Perception and Its Disorders; Res. Pub. Ass. Res. Nerv. Ment. Dis. 48*, Williams and Wilkins, Baltimore, pp. 123-138.
22. Strawson, P.: 1958 'Persons' in H. Feigl, M. Scriven and G. Maxwell (eds.), *Concepts, Theories and the Mind-Body Problem, Minnesota Studies in the Philosophy of Science*, vol. II, pp. 330-353. University of Minnesota Press, Minneapolis.
23. Strawson, P.: 1959, *Individuals*, Methuen, London.

HUBERT DREYFUS

THE MISLEADING MEDIATION OF THE MENTAL

At first when I was asked to make one unified response to the two papers you have just heard, I could see no common theme at all between a discussion of the philosophical problems concerning brain representations and their relation to the external world, on the one hand, and the philosophical problem of the identity of split brain patients on the other. In fact, there seemed to be no problem at all, let alone a common one. After all, we all have brains which are affected by energy in the physical universe, and we deal with things in the everyday world, but these levels of description seem so unrelated as to leave no space for a problem. In the same way, the brain and the person are so different it should be no surprise that a patient whose brain hemispheres have been surgically severed can function in the everyday world like a normal integrated person. Why should anyone suppose that brain representations cut a person off from the world, or that separate brain functions going on in one head cut a person off from himself?

But each speaker does seem to be grappling with some alleged philosophical difficulty raised by neurophysiology. Dr. Pribram is worried about how the brain processes are *projected* into the external world; Professor Grene, on the other hand, suggests that split brain research calls into question our notion of the *unity* of persons. In each case the problem arises when, between the brain and the world or between the brain and the person, one interposes a familiar Cartesian character: the mind. Then Pribram's problem becomes: If the representations of the external world in the brain are identical with states in the *mind*, how are these mental representations projected out into the external world? And Grene's problem becomes: If there are two *minds* occupying the same body, how can we speak of that body as one *person*? So there is a common theme after all: In each case the interpolation of a supposedly inner, private, unified mind raises philosophical difficulties when we try to understand the relation of the brain to the person or to the everyday world.

Professor Pribram takes his problem seriously and proposes a solution,

S. F. Spicker and H. T. Engelhardt, Jr. (eds.), Philosophical Dimensions of the Neuro-Medical Sciences, 131–140.

whereas Professor Grene wants to use her apparent problem to cure us of some powerful misconceptions. I will follow Professor Grene in drawing a common moral: In the light of recent advances in neurophysiology, the old Cartesian distinction between the private, inner, unified mind and the public, outer, pluralized world, which has served philosophy, psychology, and physiology for four centuries, has become more of a hindrance than a help. *Mind* has mattered too much. It gets in the way whenever we want to understand the relation of the brain to the rest of reality.

Thus there will be a common strategy in my comments on the two papers we have just heard. Criticizing the remarks of Dr. Pribram and radicalizing those of Professor Grene, I will seek to show in each case how their philosophical problems arise from the misleading mediation of the mental, and that these problems, being misconceived, cannot be solved, but can be dissolved, when we come to realize that the mental does not have the role and the importance that has been attributed to it since Descartes.

In my remarks I will be drawing upon the insights of two influential 20th-century philosophers, Heidegger and Wittgenstein, who argued, each in his own way, that although people no doubt have private feelings and experiences, and human beings are often conscious, the notion of a conscious subject underlying experiences and mediating between the brain and the world does not correctly conceptualize our everyday experience.

It is hard to get a clear focus on Dr. Pribram's paper. I am not sure what constructivism is, and it does not help much to have it situated with respect to mentalism, reductionism, emergentism, interventionalism, dualism, behaviorism, physicalism, subjectivism, and three kinds of realism: naive, critical, and inner-biological-constructional. Still, there does seem to be some sort of disagreement between Pribram and J.J. Gibson about our perceptual relation to the world, and this disagreement does not seem to be about the facts. They both seem to accept the same experimental results and both agree, I presume, that the brain plays some role in perception. The dispute, then, must be a conceptual one. On this conceptual level, Gibson with his notion of direct access to the world must be absolutely right, and I will try to explain and defend what he is getting at.

First, there seems to be a general area of agreement. Pribram sums up Gibson's position well when he says:

> As an ecological theorist... Gibson recognizes the importance of the organism in determining what is afforded. He details especially the role of movement and the

temporal organization of the organism-environment relationship which results. Still, that organization does *not* consist of the construction of percepts from their elements; rather the process is one of responding to the invariances in that relationship. Thus perceptual learning involves progressive differentiation of such invariances, not the association of sensory elements. (Pribram, pp. 104-5).

Pribram then states his disagreement:

The problem for me has been that I agree with all of the positive contributions to conceptualization which Gibson has made, yet find myself in disagreement with his negative views (such as that on 'images') and his ultimate philosophical position. If indeed the organism plays such a major role in the theory of ecological perception, does not this entail a constructional position? Gibson's answer is no, but perhaps this is due to the fact that he (in company with so many other psychologists) is basically uninterested in what goes on inside the organism. (Pribram, p. 105)

It seems at first as if the issue is whether constructive mental activity employs *images* as *elements* out of which a picture of the world is constructed. But Pribram is careful to explain that he does not hold the view that there is some point by point representation of elements of perceptual and motor information stored in the brain. "... it must be clearly kept in mind that the perceptual image, just as the motor image, is more akin to a computation than to a photograph" (Pribram, p. 108). I am not able to follow the neurological details of this argument, but the evidence that the brain computes mathematical relationships which characterize invariants in motion and perception, and the suggestion that this information might be processed holographically, seems plausible and interesting, and not something, as far as I can see, which Gibson would have to deny.

The issue, however, lies elsewhere, viz., in the relation of whatever is going on in the brain to the shared world of everyday life. This is a question that has concerned philosophers at least since Descartes; and Pribram, in spite of his modern account of what is going on in the brain, shows himself a conservative when it comes to conceptualizing the significance of this research. The crucial step comes when he moves from the physical level of speaking of a holographic representation of the external world, to thinking of this representation as a 'perceptual image' (Pribram, p. 109), i.e., when he moves from a *physical* representation in the *brain* to a *perceptual* representation in the *mind*. This way of posing the problem immediately raises a traditional but surely hopeless question:

How do the perceptual and motor representations in our mind come to be projected so as to be experienced as external to us, out there in the public world?

That this question is taken as intelligible and inevitable is a sign of the real debate between Pribram and Gibson. Gibson's view, I take it, is that what we directly perceive when the affordances of the organism mesh with the information in the environment *is* the real world. Pribram agrees that there is no place for images or elements in this view, but what he does not see is that there is no place for any sort of *internal mental representations* on this view and thus no place for the *projection* of such 'sensory events' either.

Pribram seems to think that the 'apparent externalness' of the world can be 'constructed' from such internal representations if the latter are somehow holographic.

> The holonomic model also accounts for the 'directness' of perception: holographic images are not located at the holographic plane but in front or beyond it, away from the constructional apparatus and more into the apparently 'real,' consensually validated external world. (Pribram, p. 109)

The analogy trades on the fact that a hologram at one location can duplicate by interference a light ray pattern that an object at another location would have reflected – hence, the former produces an illusion of the latter. But the analogy is absurd: what could a mental (or neurophysiological?) pattern, holographic or otherwise, be *duplicating* to produce the 'appearance' or illusion of externality? Perception of an object does not *seem* to be *just like* perception of an object – that is what it *is*.

Gibson's view seems implausible to Pribram because he takes it to deny that energy impinges on the organism and causes processes in the brain. But this is not what is being denied. What is being denied, I take it, and what should be denied, is that whatever takes place in the brain as the last step in a causal series leading in from the physical world can be viewed as a mental state, which then needs to be projected back out again.[1]

Pribram attempts to make his position plausible by appealing to the old Cartesian example of the phantom limb. "... evidence for projection comes from the clinic. An amputated leg can still be perceived as a phantom for years after it has been severed..." (Pribram, p. 108). In the phantom limb case, however, it is not clear why the phantom limb should be thought of as a projected mental state (although Descartes certainly thought of it as such), since there is no evidence that it is first felt in the mind and then projected outwards. In fact, it is probably impossible to get an example of such a confused notion as a projected mental event. Pribram tries by citing the work of Von Bekesy, which shows how points of stimulation on the surface of a subject's arms can be made to seem to be in the space between the arms. Here it is clearly a feeling which is projected, but one which is *already* outside the mind, in the body, and is then projected into the space outside the body. The same thing happens, Pribram notes, when one gets used to a probe and feels *through* it rather than feeling *it* in one's hand. If, *per impossible*, one combines the notion of the phantom limb, in which a state in the brain gives rise to an experience of an illusory object in space, with the projection of sensations beyond the surface of the body, one would get the incoherent Cartesian picture of sensations *in the mind*, projected outside the mind to give the (illusory) impression that there is a body and an external world. Thus, Pribram is led from a consideration of pathological cases and cases of learning manual skills, to the incredible epistemological claim that even veridical perception consists of the projection of inner states.

At this point the common sense assumption that we directly perceive the external world has to be saved. Pribram holds that mental states which lead to the reliable prediction of other mental states, or which are 'consensually verified' can be said to be veridical. "When what we perceive is validated through other senses or other knowledge... we claim that perception to be veridical" (Pribram, pp. 108-9).

But by calling those percepts which are consensually verified 'direct' perceptions, and so appearing to agree with Gibson, Pribram only obscures the issue. In fact, Pribram is a pure Cartesian. There is, on the one hand, the physical world which affects the brain. This gives rise to mental states or percepts. These are projected back 'out there' and in some cases they turn out to be veridical, i.e., they correspond to the matter in motion which is really out there and affected the brain in the first place. In such cases the common man (and 'naive' psychologists like Gibson) have the impression that they directly perceive the external world, but illusions such as the phantom limb reveal their mistake:

> Gibson and his followers are correct, perception is direct. They are wrong if and when they think that this means that... the percept invariably and directly [sic] gives evidence of the physical organization that gives rise to perception. (Pribram, p. 109)

This position is riddled with contradictions which have been pointed out by generations of philosophers. Pribram only further obscures the picture by agreeing with these anti-Cartesians who reject the notion of mental states: asserting with Ryle, for example, that the mind is simply the sum of mindings (Pribram, p. 99) and concluding his paper with the slogan that the mental is "extrinsic" (Pribram, p. 109), while he at the same time asserts: "Our job... is to determine which brain organizations are responsible for which mental states and processes" (Pribram, p. 100). All this confusion stems from the simple assumption that a theory of perception must relate the brain to the world via the mediation of the mental.

To get out of this conceptual trap, Pribram would have to disentangle his philosophy from his science. In the scientific study of perception it makes good sense to investigate illusions as a clue to the role of physical energy impinging on the brain. But as a philosopher one must remember that all of these investigations take place in a shared world in which we are surrounded by things and people external to us, not in our brains nor in our minds. Phenomenologically, we are in a public world, where we encounter things and people and occasionally make mistakes about them which we call illusions. These illusions only make sense on a tacit background of involved activity concerning which the question cannot even arise as to whether it is 'in here' or 'out there.' To be true to this phenomenon we must radically distinguish the *physical* level of the interaction of external energy and internal brain states and processes from the *phenomenological* level of persons acting in the world. There is no place in this picture for a third level of mental states or processes shoved in between.

Marjorie Grene alerts us to the danger of interposing a mind between the brain and reality. She ingeniously uses the latest finding in neurophysiology to call into question the notion of the unified subject with its private experiences. In this she shows both the way out and how hard it is to follow. I will try to relate her observations directly to Pribram's problem and then show that, although her subordination of the mental to the cultural is on the right track, her statement of this position is not radical enough to avoid Pribram's difficulties. We will need to purge a last residue of Cartesianism in her account before we can accept her conclusions.

Professor Grene starts right where our critique of Pribram left off, reminding us that "the myth of a unitary consciousness which somehow 'contains' [sensations]... has led philosophers, psychologists, and some neurophysiologists into a tangle of arguments and counterarguments that have plagued and, it seems, continue to plague us" (Grene, pp. 113-14).

Just as Pribram, by taking pathological brain functions as the norm, is drawn into accepting the mediation of the mental and is thus led to question whether a mind can have direct access to our world, some philosophers and neurophysiologists have asked the equally Cartesian question whether the familiar phenomena of differentiated left/right functions in split brain patients show that the patient has two minds which have no direct access to each other.

Professor Grene offers a healthy corrective to such strange speculation by reminding us that the patient is a whole person who takes himself to be one person and is so taken by his doctors and friends. She also correctly diagnoses the source of this philosophical problem in the view that to be a person is to be a unified consciousness, rather than a publicly recognized body with a more or less integrated history of meaningful behavior. She points out that the notion of the unity of consciousness has to be thought through in a new way and offers a promising proposal.

> ... we can consider our various ways of making sense of the things around us (1) not primarily in terms of 'minds,' but of 'persons,' and (2) not in terms of a simple 'unity of consciousness' as the sole criterion of personal identity, but rather in terms of a narrative model of the person which is grounded both in our bodily nature and in our cultural and historical reality. (Grene, p. 114)

However, when Professor Grene begins to work out this program in some detail, elements of the rejected Cartesian mentalism gradually creep back in. Indeed, she is almost pulled into Pribram's quicksand when she talks of "the realistic thrust... of my perceptions" (Grene, p. 118). What are these perceptions which need a "realistic thrust"? One can hear Pribram saying, "Yes, just as I said, the percept is in the mind and in order to be taken as real it requires projection and consensual validation." Indeed, once one introduces 'perceptions' the mental is slipped back in between the perceiver and the perceived object. And this is no momentary slip. For Professor Grene goes on to say that the coping that we usually do "is not only brain-mediated, but *percept*-mediated as well" (Grene, pp. 119-20; my italics).

This observation leads quite naturally to a sympathetic presentation of Polanyi's views which "most of his readers have failed to understand" (Grene, p. 124) and which, it turns out, are not so different from Pribram's incomprehensible views on projection. Polanyi draws a distinction between the subjective and personal – a dangerous move – then, using the same probe example Pribram used, argues that a tacit awareness of subjective perceptions is "transformed into a sense of... touching the objects we are exploring. This is how an interpretative effort transposes meaningless feelings into meaningful ones, and *places these at some distance from the original feeling.*"[2]

Once he has slipped in the subjective between brain and object, Polanyi, like Pribram, tries to make this position plausible by identifying the mental representation with a brain state:

Modern philosophers have argued that perception does not involve projection, since we are not previously aware of the internal processes which we are supposed to have projected into the qualities of things perceived. But we have now established that projection of this very kind is present in various instances of tacit knowing. Moreover, the fact that we do not originally sense the internal processes in themselves now appears irrelevant. We may venture, therefore, to extend the scope of tacit knowing to include mental traces in the cortex of the nervous system ([2], p. 15).

Professor Grene's allegiance to Heideggerian common sense keeps her from going this far:

To revert to feeling, or the subjective, or consciousness as the medium of our inquiry would be to fall back into the very traps set for us by the past three hundred years of philosophical thought. (Grene, p. 124)

But Polanyi has simply drawn out the implications of the introduction of some sort of subjective mediation.

It may seem needlessly drastic to sever the conceptual connection between consciousness and personhood, but Professor Grene is right that if we do not, we can be led into absurdities. An illuminating example of the strange conclusions which follow when we identify consciousness and person can be found in the literature devoted to the question whether in split-brain patients the right hemisphere is conscious. A person with only right hemisphere functions carries on integrated, goal directed activity and responds to language, but has no use of language. The question then arises: is the right hemisphere conscious?

In the Cartesian tradition the only alternatives seem to be that either it is conscious or it is functioning like a machine. Confronted with these alternatives, Sperry, for example, accepts the view that in a split-brain patient each hemisphere is conscious.

Instead of the normally unified single stream of consciousness, these patients behave in many ways as if they have two independent streams of *conscious awareness*, one in each hemisphere, each of which is cut off from and out of contact with the *mental experiences* of the other. In other words, each hemisphere seems to have its own *separate and private sensations*; its own perceptions; its own concepts; and its own impulses to act, with related volitional, cognitive, and learning experiences [4].

But if, accepting the mediation of the mental, we define a person as a self-conscious, unified stream of conscious experiences, we are driven step by step to the philosophical question raised by Roland Puccetti: Am I the person whose conscious unity is rooted in left brain information-processing and right hand motor control; or am I the person whose consciousness is based in right brain activity and subordinate left hand control [3]? And from there to Puccetti's logical but unacceptable conclusion that each of our bodies contains two persons.

Thomas Nagel in an article, 'Brain Bisection and the Unity of Consciousness,' sees that to resist the view that each of us is two minds, he must give up the notion of a unified conscious subject of experiences. But because he identifies the person with the conscious subject, he feels he has to give up the notion of a unified *person* as well. "It is the idea of a *single* person, a single subject of experience and action, that is in difficulties" [1]. "It is possible that the ordinary simple idea of a single person will come to seem quaint someday when the complexities of the human control system become clearer, and we become less certain that there is anything that we are *one* of." ([1], p. 411)

Here again the mediation of the mental, plus brain pathology, seem to force us to an impossible choice: If being a person is mediated by a conscious, unified subject, either each of us is two people, or there are no persons at all. This argument takes the form of an antinomy, since each position is argued for by showing the absurdity of what seems to be the only alternative.

Professor Grene suggests a way out of this impasse by viewing mental existence, not as the *basis* of personhood, but rather as one of its *by-products:*

> ... mental existence can be achieved... only through the complex interplay of the person's bodily beings and [cultural] artifacts... (Grene, p. 120)

This view of mind as derivative lays the basis for her admirable conclusion:

> Only the historical span of this lived body-in-its-world can identify a person as this person and no other. The loosely but subtly interwoven structures of the human brain provide one set of necessary conditions for such an existence. The more or less loosely but subtly interwoven structures of the human world provide another set, equally necessary. 'Mind,' if you want to call it so, develops in the tension between the two. (Grene, p. 121)

Before I can fully subscribe to this conclusion, however, I have to point out and exorcise one last conceptual wobble. This time the trouble comes not from an uncritical adherence to Polanyi, but from accepting the terminology of Merleau-Ponty. After playing down feeling as fleeting, and as derivative, Professor Grene states her final position in terms of a set of ambiguities: "the unity of being-in-the-world, always ambiguously psychophysical, always ambiguously public-private,... provides the ground of personal identity" (Grene, p. 121).

This formula is certainly a great improvement on Descartes' stress on the priority of the private and the mental, but it suggests that the private and the public are now to be put on *equal* footing, and the 'psycho' in psycho-physical still suggests private, inner, mental states. A later formulation in which the personal is seen as "at once mental and bodily" (Grene, p. 125), betrays quite clearly this latent Cartesianism.

I would want to make clear in any such formulation that the public is more basic than the private, and that the 'psycho' refers to public, meaningful behavior, not mental states. This leads to my final suggestion, which is, that since words like 'private,' 'subject,' 'consciousness,' and 'mental' are thoroughly corrupted by their Cartesian association, it might be best to follow Heidegger and introduce all new terminology when we speak of human being-in-the-world.

University of California,
Berkeley, California

NOTES

[1] For a more detailed analysis of this confusion and its importance in modern cognitive psychology, see, Dreyfus, H.: 1972, *What Computers Can't Do*, Harper and Row, New York, ch. 4 and Dreyfus, H. and J. Haugeland: 1974, 'The Computer as a Mistaken Model of the Mind', in S.C. Brown (ed.), *Philosophy of Psychology*, Brown and Noble, Scranton, Pa.

[2] Polanyi, M.: 1966, *The Tacit Dimension,* Doubleday & Company, New York, pp. 12 and 13. My italics. The same relapse into Cartesian thinking can be seen when, after introducing the important notion of indwelling, which in effect stresses the public, cultural status of social norms, Polanyi then reverts to speaking of the acceptance of moral teachings as their 'interiorization' (p. 17). Norms, like perceptual objects, are in no way interior, but are embodied in the world in which I perceive and act.

BIBLIOGRAPHY

1. Nagel, T.: 1971, 'Brian Bisection and the Unity of Consciousness', *Synthese* **22**, 396.
2. Polanyi, M.: 1966, *The Tacit Dimension,* Doubleday & Company, New York.
3. Puccetti, R.: 1973, 'Brain Bisection and Personal Identity', *British Journal of Philosophy of Science* **24**, 353.
4. Sperry, R.W.: 1968, 'Hemisphere Deconnection and Unity in Conscious Awareness', *American Psychologist* **23**, 724.

SECTION IV

THE CAUSAL ASPECT OF THE PSYCHO-PHYSICAL PROBLEM: IMPLICATIONS FOR NEURO-MEDICINE

HANS JONAS

ON THE POWER OR IMPOTENCE OF SUBJECTIVITY

I

Subjectivity exists. It either is what it claims to be, or it enacts a stage play behind which another type of happening hides. In the first case, its testimony – e.g., that I raise my arm because I will it – is credible at face value; in the second case, it is deceptive, a mere disguise of neurophysiological processes, which parade in the fancy-dress of will but lift the arm without will or the cooperation of will, i.e., they do so irrespective of the presence of a 'willing' sensation. The standpoint which grants the psyche its effectiveness stays in agreement with its self-testimony and needs no further reasons; the standpoint which denies its claims must have special reasons for doing so. These reasons, whatever their strength, cannot silence the disputed testimony itself, and thus the natural standpoint, constantly nourished by the subjective evidence, is never abolished in fact. But deprived of the privilege of naiveté, once the question of credibility is raised, it must defend itself against those reasons, since at the serious suspicion of an illusion even its irresistibility ceases to count in its favor. On the other hand, the suspicion must indeed be serious. Thus one must first examine the reasons which here contest the validity of immediate evidence and put naiveté in the role of a theatre-goer who takes the play on the stage for reality.

Clearly, only the strongest reasons count when it is a question of condemning all mental life to the status of illusion. I can see two such reasons and confine myself to them, ignoring all weaker ones. They are: (1) that any action of mind on matter is *incompatible* with the immanent completeness of physical determination, i.e., that the latter does not tolerate such an interference from outside: this I call the 'incompatibility argument'; and (2) that the mental as such is also *incapable* of intervention, being nothing but a unilaterally dependent concomitant of physical events and lacking any force of its own: this I call the 'epiphenomenon argument.'

S. F. Spicker and H. T. Engelhardt, Jr. (eds.), Philosophical Dimensions of the Neuro-Medical Sciences, 143–161.

The first argues from the nature of the physical, the second from the nature of the psychical. How strong is either?

II. THE INCOMPATIBILITY ARGUMENT

1. The proposition that the context of physical determination is 'closed' and does not tolerate the intrusion of nonphysical causes follows from the rule of the laws of nature, especially the constancy laws, which would each time be violated when a causal quantity with no physical predecessor were added to the given sum or subtracted from it with no physical successor. One or the other would happen whenever in the acting of living subjects the further course of events differs from what it would be without the intervention of the psychical factor, i.e., by the corporeal mechanics alone. In the physical reckoning, the intervention would amount to something's emerging from nothing or vanishing into nothing, and this is excluded by the constancy rule. Ergo: there cannot be an influence of the mental (nonphysical) upon the physical; ergo: things proceed exclusively according to the physical *concatenatio causarum;* ergo: the addition of the psychical (subjective) dimension in living beings is gratuitous and redundant for the course of events; ergo: the consciousness of aims, etc. (feelings of willing and acting), is but a deceptive imagery for the causal working of the bodily mechanism – a deception not even excused by a purpose, since *ex hypothesi* the self-sufficiency of the masquerading facts needs no such help: a purposeless deceit of purpose.

2. How sound is the argument? We observe that it invokes not simply the validity of the constancy laws, which is to be taken as inductively proven, but their *unconditional* validity, i.e., one impervious to exceptions, and this, of course, lies beyond inductive proof. Inviolability on principle pertains to the logical nature of mathematical, not of factual, rules: for the latter, it is merely postulated by us for the sake of the idea of lawfulness. The postulate originates in an idealization and expresses an ideal. The incompatibility, therefore, which the argument states, is of the type that "what must not be, cannot be." The force of the "must not" is proportional to the theoretical dignity of the ideal from which it issues. But since we, and not logical necessity, have invested it with that dignity, we can also reconsider and, if

need be, modify it, provided we stay in agreement with observable facts.

3. Moreover, an incompatibility of the nonlogical sort we are faced with designates as such no more than a difficulty of thought and leaves open *what* must be revised for its resolution: the concept of that which is to conform to a norm or the norm to which it is to conform or both? To decide this question, we must compare the strength of evidence which both have on their side but also the *consequences* from any one side's yielding to the other: in our case, by asking what becomes of 'nature' if her causal purity, or her 'exactness' is adulterated, and what of 'mind' if its effective power is denied. The scales are tipped by which sacrifice is theoretically more insufferable, i.e., more devastating for the side that is to make it. It is a question of the relative price of compatibility, if a price is to be paid. The contention I am going to argue is that on the side of 'nature' (where the all-or-nothing logic is inappropriate), the required concession in deterministic rigor would not be devastating for its scientific concept; whereas the concession asked for compatibility's sake from the mental side, viz., the forfeiture of causal force, destroys its concept completely and even drags the favored physical side down into its own ruin, leaving a caricature of 'nature' as such.

4. The seriousness of the problem lies in the challenge of materialist science to inner experience, which has on its side immediate self-certainty but no systematic predictive science; whereas natural science, with no immediate evidence for its generic ideal of objects, can produce constant heuristic confirmation for it in the systematization of phenomena. Because of this tested verification, a conflict with the ideal is a serious problem. But in the psychophysical incompatibility verdict the ideal is granted more than it is entitled to by its heuristic yield (its only evidence) and more than is necessary to preserve its rightful epistemological position. Unconditional determinism has always pretended a greater knowledge of nature than we possess and ever can possess. The holding of the constancy laws admits of degrees of rigor in detail. There is no a priori certainty that what holds for the whole also holds for all its parts down to the smallest; and what for the end result, also for all intermediate links; and what for measurable time intervals, also for each instant. The identical validity for every part and every instant presupposes an exactitude of nature which precludes any

'more or less' and 'approximately' and makes nature as pure as mathematics is. From mathematics it was that the idea of such an absolute exactness was borrowed, together with the homogeneity of whole and parts required for its application. Vis-à-vis nature it cannot be more than a postulate with which the order-loving human mind perhaps hits the truth of nature, perhaps overtaxes it.

III. THE EPIPHENOMENALIST ARGUMENT

Leaving the case of 'nature' and its alleged causal closure in abeyance, we pass from the argument that the physical does not tolerate psychical interference to the converse one that it need not fear it since the psychical (the 'subjective') is devoid of all causal force. The most complete expression of this position is the 'epiphenomenon' thesis, whose reasoning runs somewhat like this:

1. There is matter without mind but not mind without matter. The first is demonstrated by all lifeless nature plus a good part of living nature; the second by the fact that all 'mind' (here the blanket title for subjectivity of every kind and degree) appears only in conjunction with certain organizations of matter – organisms, nerves, brains – and that no example of bodyless mind is known. This observation suggests that matter has independent and primary being, mind only secondary being derivative from it.

2. Experience further teaches that matter in these forms of organization not only provides the necessary, enabling basis or precondition for mind's *existence*, and not only is the originating cause of that existence, but is also the continually determining cause for its *working* and all its changing contents – thus its necessary and sufficient cause in every respect. This is demonstrable for sense perception and certain feelings and emotions, and can hence be extended to thought processes for which it is not yet demonstrated: all these inner, subjective phenomena are wholly the effect of physical causes. But note that from this causation they do not gain a causality of their own in continuation of the former (as do effects in general), since they are nothing but an expression of what happens in the

physical substratum: mere expression cannot influence what it expresses, nor even itself, since then it would cease to be mere expression. Thus the 'induced' subjective appearance cannot emancipate itself in either sense: it can as little play within itself as it can react on the 'inducing' substratum. Its pure "as if"-pretense of both, which it displays to itself, has only the entertainment value of an illusion.

3. Joined to the impermeable causal completeness of the physical substratum previously argued, the epiphenomenon status of mind then compels the conclusion that in the case, e.g., of raising my arm, *only* the objective, neuromuscular explanation (or description) is correct, while the subjective one – in terms of will and intention – is a nonauthentic symbolic transcription for it. Neurology, communication theory, and cybernetics are at work to implement this postulate – still largely empty with regard to the higher functions of consciousness – by concrete mechanistic explanations or model constructions for increasingly complex cerebral processes. The initial successes justify the expectation, so the protagonists say, that no barrier of principle will stop progress on this road. For the rest, they challenge the 'spiritualists' to show how to represent theoretically a moving of physical entities by nonphysical movers and not play havoc in the attempt with the laws of physics.

4. Among the subjective phenomena there are the so-called 'ends' or 'purposes.' For them, too, it must hold that their subjective presence has no possible influence on the course of things. Rather, as already their presence only mirrors an objective state of the substratum, so also the putative acting '*because* of them' ensues in truth *from* the same material conditions of which this appearance itself had been a symbol. The counterwitness of subjectivity on this point is not to be accepted. Socrates, therefore, was not right when, in the famous discourse of his last night, he rejected the corporeal explanation of his sitting there and awaiting the hemlock and proclaimed the explanation in terms of mind – his ideas of right and duty – as the only true one. Neither are those right who view the two alternatives as complementary, as equivalent and interchangeable aspects of the same reality. Only the Ionian naturalists had the right view, for only the physical description explains what physically happens.

5. The test is here, as already Descartes had laid down as a rule, the possibility to *simulate* behavior by mechanical devices (automata), and this possibility has since been extended right into the mental realm, which Descartes still deemed exempt from such simulation. But then there also holds the further Cartesian principle that what is perfectly imitated is thereby no longer simulated but duplicated, and that then the duplicate discloses the *nature* of the original, to which no other principles of operation must be attributed than those needed for its duplication. Even imperfect imitation, if it is merely a matter of degree, establishes the perfect imitation as theoretically possible. Thus, if intelligent purposive behavior *can* be simulated in some simple forms, the more complex ones are *in principle* covered by the feat. Then the *ideas* of purpose, and other subjective data, which in the original concur with the behavior, are redundant for its actual performance and thus have no role in it. Thus the newly discovered possibility of imitating mind by purely corporeal means strengthens the speculative hypothesis of epiphenomenalism about the impotence of mind in general and the functional otiosity of psychological purpose in particular: like all consciousness which attributes authorship to itself, teleological belief has the character of a purely putative, operatively redundant and inexplicable "as if."

So far the materialist-epiphenomenalist argument. My discussion will focus on the *causal* aspect, which is the hub of the argument, and in its spirit I open the debate with an incompatibility argument of my own. Epiphenomenalism makes matter the cause of mind and mind the cause of nothing. But causal zero-value is compatible with nothing adhering to matter; and in particular it runs plainly counter to the idea of causal dependency itself that something dependent should be an end only (effect only) and not also in its turn a beginning (a cause) in the chain of determination.

IV. 'EPIPHENOMENON': WHAT DOES THE CONCEPT ENTAIL?

1. Before pursuing this line of thought, let us take a closer look at the concept of 'epiphenomenon' as such. It says in general that the subjective or the psychical or the mental is the concomitant of certain physical occur-

rences in brains. The 'concomitance' is one-sided, not reciprocal: the physical processes, as the primary reality, are autonomous; their secondary psychical expression is totally heteronomous or a mere product of another. The presence of the product makes no difference to the history of the producing 'other'; neither does the feat of production itself, which as it were happens 'behind its back,' on the basis of it but without a contribution from it. The product is a *by*-product of the intraphysical producing, with no expenditure deflected to its production, which thus is not a transitive act of the physical base but merely the 'appearance' of its immanent functioning. This functioning goes on as it would anyway, whether or not it had occasioned such an accompaniment. Thus the occasioning of it is *causaliter* a 'creation from nothing,' since nothing was causally spent on it. Otherwise, the epiphenomenon hypothesis were useless, since the quantity expended would physically have disappeared and its converted psychical succession were no longer quantifiable – precisely what was to be avoided for the sake of the constancy principle. *The soul's causation from nothing is the first ontological riddle which the epiphenomenon theorem braves in deference to physics, in which otherwise never a thing is supposed to arise from nothing.*

2. What thus has been produced from nothing must also remain a nothing, causally speaking. Just as the occasioning of the 'accompaniment' must cost the occasion nothing, so also its being there must not change the happening which it accompanies. This indeed is the first concern of the epiphenomenon thesis: the monocausality of matter, to be safe from interference, demands the *impotence* of mind in the first place, then its cost-free generation in the second, if it is also to come *from* matter. Admittedly, the construction becomes at least logically consistent thereby: only that which is made from a causal null can also remain a causal null, having inherited no force from its cause. And yet it is something itself and not nothing, its being there clearly different from its not being there. The appearing of consciousness adds something to the composition of reality by which it becomes different – descriptively, but not dynamically: it 'is,' but nothing follows from it for the rest of things, nothing propagates from it into their further course – the rest of reality remains what it would also be without this mirage in its midst. That which is endowed with consciousness does not behave as it does because it is conscious but is conscious because it behaves as it does, i.e., as its physical make-up makes it behave. Accordingly, the

concept of consciousness is negatively determined thus: a something for itself and yet a nothing for other things, a being without consequence, a noneffective fact. *That something, itself a consequence (an effect), be barren of consequences, is the second ontological riddle which the epiphenomenon theorem braves in deference to physics, in which otherwise nothing is supposed to remain without consequences.*

3. The no-consequence rule applies to the mental in two directions: outward toward the physical sphere, as just explained, but also immanently toward its own continuation. I.e., the impotence must, besides that of determining the body in action, also include that of self-determination of thinking in thought. Otherwise, mind would cease to be epiphenomenon and could go its own ways, on which the concordance with the body events might be lost. In that case, the illusion would explode when the mental sequence reaches the point of action, i.e., of intentional determination of the body, and the impotence would be revealed: I will one thing and my arm does the other. But it is precisely the appearance of *power* which the impotence-thesis is designed to 'save.' True appearance of the impotence would falsify the theory of impotence, which is just a theory of the deceit and not of the truth of consciousness. It is essential for the impotence-thesis that the impotence remain hidden behind the appearance of power. To ensure this, the impotence must be indivisible: only with equal inward impotence can the outward impotence remain hidden. Internal power with external impotence would result in a consciousness at odds with the world. But the theory was devised to show a consciousness united with the world – first of all with the body. Since only the world, viz., matter, has genuine reality, the unity is nothing but the negation of *any* self-reality of the other side, therefore of internal as well as external causality.

This is just what the concept of epiphenomenon provides for, by denoting something which each moment originates anew from the basis, and whose continuation, therefore, is not its own but that of the basis. As, in a motion picture, the next phase of a movement seen on the screen does not originate from the preceding phase, as it appears to do, but independently of it from the projector that emits both, and thus on the screen, contrary to appearance, in truth nothing moves – so also the temporal successor of a 'now' of consciousness cannot come from this but like itself must come from the physical substratum of which each state of consciousness is, by definition, the epiphenomenon. Thus it 'mirrors' the progress of the substratum, while appearing as progress of itself. As in the movie, this appearance is mere illusion. The hammer crashing down on the anvil is an image sequence, not a dynamic action; the deliberation terminating in a resolution is a sign sequence, not a real bringing about: its dynamics too is illusory (the internal one, even before its external sequel when, e.g., as a seeming result of the inner sequence, a real hammer is swung).

The real dynamics which the sign sequence represents is the cerebral one, which expresses (and at the same time conceals) itself in the 'text' written by it on the page of consciousness; but the continuity of the text is nothing but the continuity of the writing of it, which from below ever anew generates each sign of which it is composed and by no means lets the text write itself. The way goes not from sign to sign but from brain state to brain state and only hence each time to its equivalent in the sign script. To stay with the movie comparison: it is the continued projection that generates the continuation of the projected image: this itself is powerless. So also the 'sign' on the 'screen' of subjectivity in relation to its successor. No thought engenders further thought, no state of mind is pregnant with the next. But since precisely this is intrinsically claimed by them, it follows that all thinking is deception already in itself, and then once more when it believes to pass outside itself into bodily action. The delusion of self-determination overarches the delusion of body-determination. Both delusions are essential to the psyche. 'Soul' *is* that very delusion or make-believe with which reality constantly deceives itself. Itself? But no, brains are not deceived. The subject? But this already is a deception as such. And why? Again no answer, not even that of *l'art pour l'art,* which fits Descartes' evil spirit, but not unintentionally fraudulent matter. The deceiver, whoever be the deceived, earns nothing from his deceit. The game is played to no player's benefit and no victim's harm. It cannot serve a purpose as such. 'Purpose' is a cipher of something that is by nature purposeless. The script in which the cipher occurs – mentality – has itself no purpose, let alone a function. Its pageantry as a whole is a mirage. *The existence of such a 'delusion in itself' is the absolute metaphysical riddle which the epiphenomenon theorem braves in deference to physics.*

V. CRITIQUE

There are other riddles, of a more logical kind, to be exposed in the concept of epiphenomenon, but those shown may suffice here. Riddles by themselves do not dispose of a theory, but they surely make it suspect. We now advance some arguments against it: (1) its internal inconsistency in that it violates the very concept of nature to which it pays homage; (2) its absurdity as a theory-destroying theory.

1. The inconsistency was mentioned before: 'Soul' has to be ineffectual so that causal law be inviolate. But from this same law the 'epiphenomenon' itself would be a singular and inexplicable exception – in coming to be at no causal cost and in existing with no causal effect. Changing roles, we must now defend the principle against its defenders and insist that *nothing* in the world is 'for free' and nothing, once there, ends with itself, i.e., leaves no issue in the world. Everything caused must itself become a cause. But the 'soul,' so materialism holds, belongs to the world as a result of matter. Then (a) matter must spend something on it, and the balance sheet of a material process must read differently, depending on whether or not it has engendered an effect in the form of consciousness. *Something* of it *must* have passed over into the effect, even if in this case we do not know the equivalences for an accounting. Something was exchanged for something. Conversely, (b) the existence of this new datum (the resultant co-presence of mind), surely different from its nonexistence, must have a share in the progress of events, since no difference of existence is dynamically neutral – even if here again we are ignorant of the manner of transfer. Only on this condition has the expenditure for its becoming not simply vanished from the world, and the balance of the whole is preserved.

The principle is simply that as little as something can arise from nothing it can vanish into nothing. Nothing is pure beginning and nothing pure ending. Only the indeterminist can admit exceptions from the rule: but the materialist argues the cause of determinism and is bound to it. Thus his 'solution' contains a real self-contradiction, viz., saving the inviolability of a principle by its violation.

2. So much for the internal critique of the *concept* of epiphenomenon. More devastating still are the *consequences* which flow from it for

everything else: for the concept of a reality that indulges in this kind of thing, for a thinking that explains itself by it, and for itself as a thought of that thinking. Here the charge is not inconsistency but absurdity.

(a) First, what sort of being would that be which brings forth, as its most elaborate performance, this vain mirage? We answer: not a merely indifferent, but a positively absurd or perverse being, and therefore unbelievable. If living behavior were nothing but a deaf-mute pantomime, performed by supremely sophisticated physical systems without enjoyment of subjectivity, it could well be termed pointless but not strictly absurd. The show becomes absurd when it accompanies itself with music *as if* its predecided paces were set by it. A lie can have a function, but not here: the mechanical needs no bribe. And yet it should sound – in will, pleasure and pain – a siren song with no one there to seduce? A song that only sings its error to itself, including the error of being the singer? Something devoid of interest in the first place, and with no room for its intercession in the second, should stage the grandiose comedy of interest, shamming a task that is not there and a power it does not have? The sheer, senseless futility of such an elaborate hoax is enough to disqualify it as a caricature of nature. He who makes nature absurd in order to circumvent one of her riddles has passed sentence on himself and not on her and has forfeited the right to speak any more of laws of nature.

(b) Even more directly than via the slander of nature has he passed judgment on himself by what his thesis says about the possible validity of any thesis whatsoever and, therefore, about the validity claim of his own. Every theory, even the most mistaken, is a tribute to the power of thought, to which in the very meaning of the theorizing act it is allowed that it can rise above the power of extramental determinations, that it can judge freely on what is given in the field of representations, that it is, first of all, capable of the *resolve* for truth, i.e., the resolve to follow the guidance of insight and not the drift of fancies. But epiphenomenalism contends the impotence of thinking and therewith its *own* inability to be independent theory. Indeed, even the extreme materialist must exempt himself *qua* thinker, so that extreme materialism as a doctrine be possible. But while even the Cretan who declares all Cretans to be liars can add, "except myself at this moment," the epiphenomenalist who has defined the *nature* of thought can not make this addition, because he too is swallowed up in the abyss of his universal verdict.

Thus we have a twofold *reductio ad absurdum*, according to the twofold question of what to think of a reality that brings forth this futile mirage and what of the attempt of this self-confessed mirage to establish a truth about that reality. Nature as an impostor on the one hand, a theory destroying itself on the other, was the outcome of the scrutiny.

3. But, so we ask at last, was this suicidal *tour de force* really necessary? Are the reasons prompting it indeed so compelling that they only leave this counsel of despair – or that of conceding defeat before an insoluble riddle? The latter, of course, would be the proper option if this were the choice, and there would be nothing shameful about its modesty and honesty. But I think one can do better. I think one can show to the upholders of physical causality that their cause admits a resolution of the dilemma that does not injure it on their own terms. To this end I take now the liberty to engage in a thought experiment which is not meant to produce a theory but merely to illustrate possible compatibility by a freely constructed logical model.

VI. TENTATIVE SOLUTION OF THE PSYCHOPHYSICAL PROBLEM

1. Let us assume: A geometrically perfect cone stands with its apex on a surface over a center of gravity which lies exactly in the prolongation of its axis, i.e., this rises exactly 'vertical' from its support. With perfect symmetry and complete absence of other forces or force-differentials the cone would stand in absolute, but absolutely unstable, equilibrium. 'Absolutely unstable' means that an infinitesimal influence would suffice to tip it over and make it fall to one particular side. Let us think it to be a giant cone: the smallest disturbance of symmetry from without or within would trigger giant consequences. These consequences in their total course, in acceleration, force of impact, mechanical and thermal effects, equalization of the initial energy gradient, would each and all be governed univocally by the laws of nature and be calculable according to them – except the accident of the direction of the whole, for which all other directions, i.e., an infinity of such, were equal candidates. As to the trigger impulse that did decide the direction, its minimal value does not enter into the measurement of the consequences with their incomparably higher order of magnitude, and for

purposes of *their* causal description the unknown starter x equals zero. To be sure, the universal physical hypothesis would postulate for this x again, infinitesimal as it be, that it was determined in accordance with the constancy laws; or generally that there was a 'sufficient reason' in the antecedents for this one direction of tilting being selected over all the equally possible ones. But operationally, the postulate has to remain empty for lack of verifiability, and it is obvious that here, at the critical balance point, there is room for sheer randomness or indeterminacy without detriment to the strict determinacy of the movement once underway. Indeed, computationally it would make no difference if for the initial x we posited a psychic, nonphysical origin: in the present case there is no reason whatever for doing so, but this may change as we come to other examples. So far the thought experiment shows that neutral interstices in the matrix of physical determination, properly located, would not interfere with the validity of physical law and the intactness of causal 'bookkeeping.'

2. Our first example was nothing but the simplest, crudest, and thus utterly unrealistic illustration of the *trigger principle*, which plays such a crucial role in higher organizations of matter, i.e., in the kingdom of life. Let us go here straight to the summit of the pyramid where we encounter the subtlest form of the 'inverted cone.' Assume we have the trigger points *A, B, C*... at the primary control centers of efferent nerve paths in the brain, corresponding to the respective possible motor commands *a, b, c*..., thus representing the 'yes of no' for the actions α, β, γ...; and further assume a physical state in which the chances of activation are equal but alternative for all (any one, but one only, being eligible), i.e., that the decision on *which* of them will be 'fired' is completely in the balance; and finally assume that for the activation, i.e., for occasioning the transition from potentiality to actuality ('triggering'), an influence of the smallest order is required: What then would be the situation physically if the 'choice' has actually fallen on *A*? Well, the ensuing transaction *a*, i.e., the neural transmission of the 'command,' and then the transaction α, i.e., its motor execution (and thereafter everything which follows from it in the external world), can in unbroken sequence of determination be described, and thus explained, according to the laws of nature (thus ideally also be predicted) – without there having to be an answer to the question of why *A* rather than *B* or *C* had been activated. Since the magnitude responsible for that has zero value

for the account of what is observable thereupon, it makes no difference for the latter's consonance with the laws of nature whether *A* or *B* or *C* has been activated: the alternatives are equally orthodox in their physical behavior, equally possible 'before' and equally deterministic 'after,' and only the decision on which of them is allowed to become operative is indeterministic *on this plane* of calculability.

Nevertheless, for this *x* too (i.e., for the initial resolution of indifference) one might in theory at least insist on a physical accounting, considering that the relative 'zero' of the triggering factor is, of course, not a real zero but *some* quantity (even, according to quantum theory, of an indivisible minimum value), and one has a right to ask whence this came and what *its* prior determination was. Two answers are possible on physical terms: Either its incidence on *A* was a purely random event in the given quantum field, or it followed deterministically from the antecedent distribution therein. Both answers are physically acceptable on principle, but both would here be in contradiction with the facts: the randomness with both the physical and mental facts, the mechanical determinism with the latter alone (as shown before), in that it would allow them no power at all.

3. There is a third alternative, which admittedly would no longer be a physical explanation: that the physical quantity required for the selection of the neuron (a sort of 'Maxwell's demon') is *generated* on the part of 'subjectivity' or 'psyche,' i.e., from beyond matter. Let's not be too frightened by the anathema of this idea to physical orthodoxy to examine its implications. It speaks of a *de novo* increment to the antecedent physical sum, the insertion of a value not accounted for from within the physical system, and in that respect a *creatio ex nihilo*. Its smallness would make it unspottable in any computing of physical factors for the macrosequence of events, but the effects of its trigger function in the unstable equilibrium of the threshold condition can be immense: war and peace, rise and fall of empires, building of cathedrals and atom bombs, greening or wasting of the world may be 'caused' by its infinitesimal contribution. All these chains of action in their macrocourse would offer to causal analysis a deterministic picture in complete accord with the constancy laws, as *also would innumerable others*: only the pre-beginning of each, which selected it from among the alternatives, would be indeterministic, i.e., not physically *but mentally* determined – just *as direct experience tells us*. From the point of

view of nature, the one 'direction' actually taken would be (as with the cone) an absolute accident, which as such is beyond verification.

4. So far, so good. But an objection is obvious: many vanishing magnitudes can add up to noticeable ones; and what we have been speaking about was not a rare event, but something occurring constantly and in countless cases, viz., with every outward action of every 'subjectivity'. Thus, even if willing, uneasily enough, to grant the occasional happening of a sufficiently small 'creation from nothing' in the context of physics, we may well balk at what our suggestion seems to amount to: that the world of life, across the whole breadth of subjectivity diffused through it, incessantly donates antientropic energy to nature, unpaid for by increased entropy elsewhere. This, apart from its repugnance to general theory, robs the hypothesis of its saving grace of being innocuous to the rule of the constancy laws, since the accumulation of the singly nonmeasurable must eventually grow into the dimension of the measurable and, hence, into visible conflict with the requirements of those laws.

5. So it would be indeed, if the case as presented up to now were complete. But its reverse has yet to be added, and when complemented by it, the hypothesis by no means stipulates a one-way flux of becoming. First we simply recall that matter was supposed to be the prior donor in supporting subjectivity as such, and so it may just receive back its original outlay in the 'influx' we discussed. More especially we must remember that – as afferent nerves correspond to efferent ones – consciousness in its overall world-relation is essentially a two-way street and not a one-way street. Action *into* the world, thus far considered alone, is based on information, an input *from* the world, i.e., ultimately on sensibility. But in every act of sensuous affection, the physical chain terminates in a mental representation = the percept in question, and this too cannot be free of causal cost: some value must have vanished from the physical ('objective') side to reappear on the mental side in the radically different form of subjectivity. If we were nothing but contemplative beings who only perceive the world, there would be constant transfer and drain from the physical order, thus the same embarrassment as before with the opposite sign; if only active beings (e.g., with an a priori intellectual knowledge of all objects of action), then there would be constant reverse transfer and inflow with *its* embarrassment. Since we are

inseparably both, and this in essential complementarity, it is not unreasonable to assume that in the physical average, outflow and inflow balance each other with reference to the whole phenomenon of subjectivity and the two opposite embarrassments cancel out.

The key to a solution of the psychophysical problem, that peculiar impasse which nothing but a philosophizing physics has created for theory, lies – so we suggest – in an age-old insight which has never been utilized in this connection: that our being *qua* subjects has this double aspect and consists of receptivity and spontaneity, sensibility and understanding, feeling and willing, suffering and acting – in brief: that it is passive and active in one. To what model construct, then, does our thought experiment lead?

VII. THE SPECULATIVE MODEL

Using a metaphor, we say: the net of causality is wide-meshed enough to let certain fish slip in and out. Or with a change of metaphor: at the 'edge' of the physical dimension, marked by such peaks of organization as brains, there is a porous wall, beyond which lies another dimension and through which an osmosis takes place in both directions, with a priority of that from the physical side. What thus physically seeps out and in is of too small a magnitude to show up quantifiably in the single case and mutually so balancing in the total as not to affect the verifiable overall working of the constancy laws. In virtue of the trigger principle, the smallness of the single input does not preclude great physical effects. Passage through the 'wall' means each time a radical transformation in kind, such that any relation of equivalency, even the very meaning of quantitative correspondence, ceases to apply. The greatest thoughts with the mightiest consequences can arise from the tiniest physical input, and the tritest just as well. What matters is that between input and output there is interposed a process of an entirely different order from the physical one. Short or long as may be the loop of the circle that passes through the mental field on the other side of the wall, it does not move by the rules of quantitative causality but by those of mental significance. 'Determined' it is too, of course, but by meaning, understanding, interest and value – in brief, according to laws of 'intentionality,' and this is what we mean by freedom. Its yield is eventually fed back into the physical sphere, where everybody can recognize it (for everybody knows that unthinking nature builds no cities), without any single physical nexus confessing to its share. With the transfer forth and back, where egress and ingress go on continually, the total balance for the physical side remains even (nothing analogous applies to the mental side), and it is the level of that balance on which natural science does its explaining. The *understanding* of the same event is done from the level of that which for the moment stands outside the balance and is 'transcendent' to it in this sense. In that understanding, the extraphysical interlude is recognized as the true origin of the physical action, though only infinitesimally its 'cause.'

For mentality, thus, the brief formula holds: generated by minima of energy, it also can regenerate minima of energy. In between, these minima are gone from the physical surface, yet have not vanished into real nothing any more than has a subterranean river; thus, they also do not emerge from a real nothing when consciousness acts back into the world. The 'in-between' itself is the realm of subjectivity and (relative) freedom.

Accordingly, the brain is an organ of freedom, but precisely on condition that it is an organ of subjectivity. To put it the other way around: Supposing a brain of the same physical constitution as the human, but without concomitant subjectivity, we contend that it would not produce the same effects in the visible world (though perhaps quite respectable ones in body control) which we know the human brain to do. That is to say, subjectivity is not a causal superfluity. No more need be said here about the difference in the 'intelligence' of machines and of the human mind. Actually it is my view that the hypothesis of a 'merely physical brain' (except in a corpse) is inadmissible: Such a physical organization *eo ipso* means in its functioning the opening up and sustaining of a psychic dimension, which then participates in the overall causality of the system with the leverage of its key position. Obviously the freedom thus established is not absolute but confined to the latitude which physical necessity itself allows it. It does allow, as we have illustrated, that the smallest force can wield the greatest power when in the given 'critical' configuration it suffices to 'tip the cone.'[1]

That there are, in recurrent readiness, situational sets of such poised 'cones' and thereby sets of options among physically equivalent possibilities, is the functional meaning of such physical organizations as brains in control of organisms. Riding on the crest of this physical organization – one of whose roles is that of amplifier – 'mind' with its immeasurably small physical input can be the initiator and determinant of physical effects in that order of magnitude in which visible behavior takes place.

Ontologically it should be noted that, according to this model, the 'beyond' of the dividing wall or membrane is not a no-man's-land which keeps its inhabitants for itself and in which they can lose themselves as in a spirit world. Just as it only lives on the continuous input from the physical side, so it feeds back into it what has gone through the transformation of subjectivity. Mind thus belongs to the one and same ontic reality as matter, only with a thoroughly different nexus of ontologically different elements within its own dimension. In other words – as the voice of the 'self' has always been telling us – the one, coherent, convertible and intercommunicating Being is not exhausted by its massively prevalent physical aspect.

The model we have constructed is admittedly crude. We need not bother with whatever refinements it is susceptible of, since it does not even claim to be 'true,' i.e., to portray what is actually the case. It is a mere play of thought, meant to illustrate that *on the 'physicalist' premises themselves* the psychophysical impasse of their making (and theirs alone) logically admits of better solutions than the wholly unacceptable one of epiphenomenalism. As a point of strategy, not of conviction, we have made maximum concession to the materialist case and its conception of the nature of matter. The truth, I suspect, would look vastly different – not only more subtle but also framed in ontological terms which would alter our very speech of 'matter' and 'mind.' For our purpose it was enough to come up with a hypothetical fiction in the conventional terms that satisfies these three requirements: to be self-consistent, to be consistent with observable facts, and to spare nature and ourselves the scandals which materialist epiphenomenalism has been shown to inflict on both: on nature the scandal of an alleged procreation of mind without causal cost and consequence, on mind the scandal of utter futility in both action and thought. We have at least recovered, even for the captives of physicalist creed, a good intellectual conscience for believing in the immediate evidence of our being.

VIII. CONCLUDING REMARK

Has any of this a bearing on medical theory or practice? I indicate one and hope for more from my commentator and others (if the gist of the preceding argument is accepted).

The concept of psychosomatic illness remains half-serious so long as a genuine causality of the 'soul,' of the purely mental, is denied, i.e., so long as the mental itself is conceived as no more than a byplay of the body. This by itself is a point of mere theory, but it may affect therapy. Therapeutically, the materialist hypothesis must always favor direct physical action – chemical or surgical – on the brain and regard psychotherapy at best as a roundabout way of physical intervention in mental disguise. If mental causality is authentic, its enlistment in the healing process becomes authentic too, while at the same time the avowed somatic condition for its coming into play will not therefore lose in attention. It remains, of course, an entirely pragmatic matter in the given case to decide from which side the intervention can most effectively start. The inextricable togetherness of both sides is axiomatic to all hypotheses we can possibly adopt.

New School for Social Research,
New York City, New York

NOTE

[1] This view of the matter would, e.g., rule out telekinesis and other spiritistic macroeffects.

STUART F. SPICKER

THE SPURIOUS PSYCHE-SOMA DISTINCTION

I. INTRODUCTION

Philosophers are fond of unveiling arguments which give the illusion of correctness, soundness, and validity, especially when others have already been persuaded by their seeming correctness. Among the informal fallacies noted by logicians is the famous *argumentum ad hominem*. The abusive form of that fallacy is committed when, we are reminded, instead of trying to disprove the truth of what is asserted, one attacks the man who makes the assertion. In this regard I find myself in an odd if not embarrassing position as commentator on Professor Hans Jonas' paper ([9], pp. 143-61), for it is neither to the point to attempt to disprove the truth of or to reaffirm what he has asserted, since he has admitted that his constructed model, which serves as a tentative solution of the psychophysical problem, "does not even claim to be true," nor does it "portray what is actually the case," since he has only engaged, by recourse to a thought experiment, in "a hypothetical fiction." Furthermore, as student and disciple of Hans Jonas, I could never wittingly condescend to any *argumentum ad hominem* even were I to think myself clever enough to make such an "argument" palatable to this auspicious audience. The *curriculum vitae* of Professor Jonas reveals (may I be permitted to say) the excellence of the man and not just his life and thought. If I may add a brief autobiographical remark, I should mention that while a graduate student at the New School for Social Research, I was nourished by his careful critique of Aristotle's *De Anima*, Plotinus' *Enneads*, Kant's first *Kritik* and, of course, through his writings, the ancient Gnostics. Indeed, in the Introduction to his *Philosophical Essays*, published as recently as last year, he remarked that "Gnosticism had been the most radical embodiment of dualism ever to have appeared on the stage of history..." ([8], p. xviii).

Consistent with his abandonment of the unsound arguments of the Gnostics, he rejected Cartesian dualism – more correctly, he rejected the

S. F. Spicker and H. T. Engelhardt, Jr. (eds.), Philosophical Dimensions of the Neuro-Medical Sciences, 163–177.

metaphysical dualism disclosed in the writings of René Descartes with its preoccupation with and "exclusive focus on mentality" ([8], p. xii) – reorienting his thought by attending to the concept of organism which, to be sure, did not entail his rejection of Descartes' entire corpus. Clearly, much of what he has argued today is designed to revisit contemporary Cartesian dualism, which has served to alienate human existence from itself; mock Nature as a mere 'caricature,' an 'impostor' and the source of a 'hoax'; degrade the realm of organism and organic life; denigrate natural science through uncritical philosophizing; and tear the conceptual fabric of the very notions of 'mind,' 'matter,' and 'Nature' by acts of intellectual violence.

To counter these contemporary forces Jonas adopts a strategy which reveals four elements: (1) to concede from the very start materialism's concept of the nature of matter; (2) to demonstrate that on the physicalist premises themselves the psychophysical impasse disclosed throughout modern and contemporary Occidental philosophy does not entail the truth of epiphenomenalism; (3) to argue that the dilemma posed by the upholders of physical causality who address the mind-body problem is resolvable; and (4) to show that his suggestion for the resolution of the psychophysical problem is consistent with the metaphysic of 'physicalism' or 'materialism,' the latter terms being univocal.

By making concession to materialism's concept of the nature of matter and offering arguments for the truth of the three propositions entailed in his strategy, Jonas hopes to satisfy three requirements: (1) to be self-consistent; (2) to be consistent with observable facts; and (3) to spare Nature and ourselves the scandals which materialist epiphenomenalism has been shown to inflict on Nature and human 'subjectivity,' 'mind,' 'psyche,' or 'inner experience,' the latter terms used univocally.

I will have little to say about the last two aims, since in the main I agree that (1) Jonas' view is consistent with observable facts and that (2) epiphenomenalism does embarrass man and insult Nature, i.e., epiphenomenalism is not coherent with respect to human experience and the human condition, nor does it adequately treat the relation that obtains between consciousness and Nature, however wrongfully one construes the subtleties of that relation. But I fail to see that Jonas' 'resolution' of the psychophysical problem is fully self-consistent – keeping in mind that with great caution he does not maintain that it is true, actually the case, nor demanding of our assent, since it is only tentatively proffered.

Returning to his strategy, then, I shall not take up in great detail the view (which I believe we share) that he concedes too much to the physicalist by allowing him to propound, unassailed, his concept of the nature of 'matter,' the 'physical,' the 'corporeal' (which for me are not terms interchangeable, *salva veritate*). We might recall that for Jonas it is "a point of strategy, not of conviction" that he made a maximum concession to the materialist case and its conception of the nature of matter. Respecting the fact that Jonas himself abstains from any commitment to physicalism, I shall not argue the case (one with which I believe he agrees) that physicalism or materialism is false, though I suspect his account of mind's causal efficacy lends credence to that view among those who are less than careful in appreciating the strict strategy he has adopted; but to indulge in expanding this point would entail a digression from the main line of the argument, which I must here forego.

I shall, therefore, attend to Jonas' solution to the dilemma posed by the mind-body problem and, by raising questions, suggest that he does not in fact resolve the dilemma – not because his view is somehow consistent with the false metaphysic of physicalism and/or entails epiphenomenalism, but simply because it is conceptually unclear and unconvincing in terms of his notions of 'cause,' 'psychic cause,' and 'significance.' I shall attempt, in closing, to carry out the charge he has set for me and point the way to the theoretical bearing his themes have for neurology and psychiatry, as well as consider their therapeutic and clinical import. In proceeding I am fully aware of the danger that awaits the disciple should he too rashly assume he has understood the master; on the other hand, one test of an authentic disciple is that he in some measure misunderstand the master.

II. MIND'S CAUSAL EFFICACY

Professor Jonas' understanding of mind and body has its roots in everyday experience; he begins by taking quite traditional and commonplace views about matter and mind. A careful perusal of his use of these concepts, later expressed as the psychophysical problem, reveals that his formulation is very much like that proffered by Descartes and received by Princess Elisabeth of Bohemia in their famous correspondence in May and June, 1643. The human body is, throughout much of his argument, quite like but not identical to an inanimate body; it has size and weight, is moved by the force of other bodies, is causally efficacious when affected by external forces, is usually in a state of equilibrium with respect to gravitational forces, but if at a given time not so, then is easily moved in other directions, eventually assuming a place at rest as a consequence of the resultant of forces acting on it. Sometimes a 'minima of energy' ([9], p. 159) is required to move a bulky and heavy body, and one can thus imagine an inverted, ancient Egyptian pyramid in balance on its apex easily toppled by a relatively small physical force ideally and operationally defined as the least effort required to 'trigger' or violate the equilibrium and balance of the pyramid.

Minds, on the other hand, do arise, to all appearances, within the physical world; they do remain, to all appearances, bound to living animal organisms; they are not, for Jonas, simulatable; they appear to be the source of purpose in the world; they are capable of 'thinking,' that term taken in the widest Cartesian sense of the term (in contrast to Descartes' restricted use of cogitating as is also found in his *Meditations on First Philosophy*). That is, minds doubt, understand, assert, deny, approve, disapprove, will, imagine, hallucinate, and perceive things and events in the physical world.

Jonas has adopted a formulation of the psychophysical problem in which it appears that mind and matter are two distinct substances of different types, each of which has its own appropriate states and properties; more importantly, given this formulation, it then follows that none of the properties or states of either body or mind can be a property or state of the other. States of consciousness (mental states) belong appropriately to the mental substance, to subjectivity, mental life, or consciousness, and not to body, matter, or physical reality. But this formulation Jonas seems to reject, for he finds it perfectly acceptable to say that not only matter or body possesses causal efficacy, but so too does mind or subjectivity. The psychical, he says, is not devoid of causal force; it can work. "Subjectivity is not a causal superfluity" ([9], p. 159). Epiphenomenalism is false precisely because for it matter is the cause of mind, whereas mind is "the cause of nothing"; for it "the psychical is devoid of all causal force" ([9], p. 146). Mind, then, for Jonas, has a great power, but it does not reveal a 'quantitative causality.' Moreover, it only takes an infinitesimal psychic cause to generate great mental *significance* as well as move human bodies or (when one wills to raise one's arm) parts of these bodies. Furthermore, the 'minima' of psychic causal efficacy is too small to quantify, he tells us, though it wields the greatest power. One is reminded here of Leibniz's *petites perceptions* (as well as the fallacy of the thesis that accompanies them), exemplified by the unperceived sound of each drop of the ocean's wave, whereas the sound of the totality of *minima perceptibilia* is heard in the mass as the roar of the noise of the sea as it strikes one when on its shore. This view leads to the incoherent and contradictory picture that we have perceptions which we do not even perceive; it is claimed that there are unnoticed perceptions or mental states not strong enough to attract our attention. Thus my first question:

As it is incoherent to maintain that there exist mental states which are supposed to exist as such below the threshold of consciousness, which remain unnoticed, in Leibniz's view, likewise is it not superfluous, if not incoherent, to maintain that subjectivity or mind generates the activity of a neuron with 'unspottable' micro force on the macro world and, in addition, is *unexperienced* by the person within whom the micro force occurs and whose arm is then raised and *experienced* as raised?

Can't one beg (along with Princess Elisabeth) for someone to tell how the human soul or mind can determine the movement of the animal spirits in the body so as to perform voluntary acts – mind being, as it is, merely a conscious [*pensante*] substance. A 'psychic cause' would seem to require contact, she pleads, as is the case when we speak of material cause or physical cause. Yet contact seems to be logically incompatible with Jonas' view of mind, for contact seems to be incompatible with a thing's being immaterial, as indeed is mind or psyche in Jonas' view. For nowhere does Jonas argue that mind is material. The fact that a mind can suffer along with and be responsive to body does not entail that we can predicate of mind *causal* efficacy, which somehow penetrates, as he fancies it, by 'osmosis' through the 'porous wall' which separates mind from body, i.e., mind from brain. The objections to this model are the same as those which Descartes' doctrine of the union of mind and body had necessarily to face when he appealed to the pineal gland as the seat of the soul, the locus of their interaction [2].

But perhaps Jonas takes the principle of 'action at a distance' as paradigmatic of mind's causal force. Descartes was, of course, already aware of this suggestion as is evidenced in his correpondence with Elisabeth, in which he suggests that people tend to confuse the notions of the soul's power to act within the body, *and* the power one body has to act within another. For him, the power of the mind to act within the body is not to be ascribed to gravity or something like it, a construct which exemplifies the principle of 'action at a distance.' Thus, for Descartes, the notion of mutual contact served as the best simile or analogue for comprehending the way that the mind moves the body. But in a subsequent letter Descartes admitted that his simile of gravity was "lame." Elisabeth replied that she "cannot see why this should convince us that a body may be impelled by something immaterial" ([3], p. 278). She simply believes that it is impossible for immaterial mind to move the material body. Like her, I confess that I can more readily allow that the mind has matter and extension than that an immaterial mind has the capacity of moving a body and being affected by it. Descartes' reply to this is, of course, that this is nothing else than to conceive the mind as united to the body.

Jonas thus promulgates an interactionism – one which even supersedes Spinoza's monistic (eternal substance) parallelism as well as all the weaker forms of dualism, including epiphenomenalism. He holds there is causal efficacy both ways; thus some mental events are caused by brain events and some brain events are caused by mental events. But, so far, adequate empirical evidence for independent psychical cause is yet to come forth, notwithstanding the fact that great advances in instrumentation yielding refined measurement of subtle brain and nervous activity is now extant. That is, as far as we can tell today, all mundane causality is physical and, in general, human bodies do not seem to offer evidence that they are immediately and causally determined by minds with which they are most intimately conjoined. Thus my second question:

Is not so-called "psychical causality" a factitious idea – a mere speculation and thus superfluous (which does not entail the view that subjectivity is superfluous)? Is it really a viable tactic to construe a chain of psychical causes, independent of a chain of physical causes, thus meeting the physicalist on his own ground, if not conceding to him the entire battlefield? Do we not have a right to call for a detailed analysis of this psychical causality when Jonas' thesis has it that this is involved in the interaction of a mind with its body?

For the most crucial issue, I submit, is not that there is or is not psychical causal force (which let us grant we might one day conceivably be compelled to acknowledge due to empirical discoveries), but rather how such a psychic, causal efficacy can affect a brain, central nervous system, and entire body with which the individual psyche is so intimately conjoined. Put quite simply, what is the nature of the process which goes by the name of psychical or 'internal causality'?

In spite of my skepticism here, Jonas' interactionism is surely superior to dualisms like epiphenomenalism and naive monisms like physicalism or idealism, since he makes body and mind parts of one *ontic* reality, that is, Nature. It is interesting to note that in his classic paper, 'Spinoza and the Theory of Organism,' published in 1965 [7], Jonas was already discontent with Spinoza's position as expressed in Part III, Proposition ii, of his *Ethics*, to wit: that interaction is false, Spinoza's position being that "The body cannot determine the mind to thought, neither can the mind determine the body to motion or rest, nor to anything else, if there be anything else" ([7], p. 62). In that marvelous paper Jonas also remarked that "none of the leading thinkers of the period down to, and including, Kant ever challenged the validity of it," i.e., the validity of Spinoza's proposition. But Jonas now holds that the insistent evidence of common experience seems to count against Spinoza's proposition. Overwhelming consensus, Jonas said then and argues now, has it that the mind, for example, mental acts of intent or will, can and do *determine* action and bodily motion. And thus the seeds were planted in 1965 for today's seminal attack on epiphenomenalism. So all the difficulties seem to hang on Spinoza's notion of *determination* and Jonas' conception of *psychical cause*; whereas Spinoza rejects interactionism, Jonas embraces it. The third question, then, I now pose:

In following Jonas' view (even as a hypothetical fiction) we pay too high a price. For we not only pay too much, we pay more than once. First, do we not concede too much to physicalism by apparently allowing the move that mind is (at least in principle) material, possessing the power of causality which insinuates itself "across" to the brain? Second, does not such a concession make identity theory, a form of naive monism, seem more plausible? For it now appears that it is not merely the case that *whenever* a physical state occurs *then* a concomitant mental state does also, but that mental states *are* physical states of the brain and CNS (a view which is patently false, for brain states are temporal and spatial, whereas mental states are not spatial, though temporal; *and*, moreover, the time of mental states is not really the time of physical processes of the brain). Third, does not this concession to physicalism encourage a return to man in the image of 'machine,' an image I believe Jonas would never concede to at any price, for it encourages one to adopt a Pinnochio view of man which Jonas rejects in the image of the "deaf mute pantomime"?

I wonder if Jonas, in the end, will not re-cognize Spinoza's ingenious theory of psychophysical parallelism as a *better* account of the relation between levels of material organization and states of mind, one which is consistent with the principle of noninteractionism? Perhaps this is the significance, albeit surreptitiously introduced at the close of his paper, of his claim that "mind belongs to one and the same ontic reality as matter." Does he not really intend to say '*ontological*', not 'ontic', here? This is, of course, Spinoza's view, one in which we are presented with a novel conception of substance as absolute and infinite, thus making mind and body not substances, since they are finite, but modifications of substance, that is, modes. Jonas' version of interactionism reveals that psychical cause, whatever it may be, if it be anything at all, must be sufficiently distinguishable from physical cause as to warrant a separation of two forms of reality, mind and body, and this violates Spinoza's psychophysical parallelism, for which there is as little substance 'body' as there is 'mind.' The language of Jonas' interactionism permits us to say that each material event has its *counterpart* in a mental event, but Spinoza's metaphysic rejects this as too disjunctive. With Spinoza the Cartesian riddle, as Jonas remarked in 1965, "disappears." Does it return to haunt us today in the form of Jonas' psychical causal determinacy and metaphysical interactionism? Do we not come full circle and return to the original riddle, obliged to recall Descartes' remark to Elisabeth (when he said that she may think he is not speaking seriously when he said) that "it is just by means of ordinary life and conversation, by abstracting and meditating and from studying things that exercise the imagination, that one learns to conceive the union of soul and body"? In sum, it may well be a propitious moment to observe that Descartes' replies to Princess Elisabeth were something less than the expression of a commitment to good intellectual conscience and logical argument; his treatment of Elisabeth being (if you will) not too high-minded, either. In our day such chauvinism quickly lends force to the cause which demands a Latin neologism to underscore a more feminine form of the informal fallacy *ad hominem*.

III. PHENOMENOLOGY AND THE PSYCHOPHYSICAL PROBLEM

In John C. Eccles' edited volume, *Brain and Conscious Experience*, E.D. Adrian remarked that the agreement among metaphysicians in rejecting dualism "has not been coupled with agreement in accepting anything else" ([4], p. 239). In anticipation of this serious criticism of the failing of philosophy and philosophers, my opening remarks included a reference to The New School for Social Research, which was not intended as merely autobiographical, for it was at the New School that Edmund Husserl's phenomenological philosophy was transplanted onto American soil, coming to fruition in the thought and teaching of the late Professors Dorion Cairns and Aron Gurwitsch. In his paper Jonas pays homage to Husserl's theory of intentionality (another name for phenomenology) when he refers to the mental field as not governed "by the rules of quantitative causality but by those of mental significance" ([9], p. 158). The mental field is "determined" not, after all, by causal force but, as he says, by "meaning, understanding" according to the laws of 'intentionality' (a term not to be confused with what the analytic philosophers call 'intention'). In brief, Jonas' telling sentence offers us a clue to the resolution of the psychophysical problem; it is now to be understood in terms of intentionality, a theory of meaning constitution described, in part, by the complex intentional founding-founded structures which reveal the strata of sense of consciousness (subjectivity) and are explicated in some detail in Husserl's *Ideen II.*[1] Husserl was well aware of the interdependence of body and mind, i.e., that organized matter or organism is the necessary condition for all neurophysiological functions, yet his phenomenological standpoint did not reduce mind to body or body to mind. The *sense* 'mind' and the *sense* 'body' are distinct as sets or systems of relations. From this perspective neurophysiology is to be understood as charged with clarifying the complexity of the brain and CNS, that is, the intricacies of the organic basis of the patient. The organic body is one way of describing the significance of mind. Thus mind and body are distinct only in thought, though inseparable in fact. In this Descartes was correct when he remarked that "the whole mind is united to the whole body" ([3], *Meditation* VI, p. 121) and not to one of its parts only. Husserl revised Descartes' substance or 'thing' theory, however, by suspending belief in the world as existing as is understood in the natural attitude, in which mind and body are conceived as substances. He interpreted them as realms of *significance*, structured as related strata of sense or meaning and made explicit by a method of 'cognitive archaeology.' This was expressed in terms of the constitution of founding-founded relations ([6], p. 55 *et passim*). The full exploration of this 'archaeology of consciousness' is, in short, Husserl's phenomenological program; it is nothing less than an explication of the structure of *significance*. Thus the model of causal connection which was presumed, in the natural attitude, to bridge the gap of mind and body, was abandoned as unefficacious. All accounts of mind and body as substances prove recalcitrant ([6], p. 82). The psychophysical problem was misposed (one can now say) when it appeared to require an account of the interaction of *things*. An adequate explication requires a radical tranformation of the question, "the problem," itself. What is needed, in part, is an account of the categories of the lived-body,[2] and to accomplish this one must have recourse to the realm of significance which Jonas signals in his reference to 'intentionality,' 'significance,' and 'meaning.' Jonas' reference to these notions suggests, then, the abandonment of Spinoza's "solution" to the psychophysical problem, as well as Descartes'. Once we abandon the notion that mind and body are substances, 'things,' we are in a better position to appreciate what we *are* as psychophysical unities (what I prefer to call 'lived-bodies' rather than simply 'organisms'). Husserl's founding-founded analyses of the structure of psychical and physical significance permit both a psychological analysis of mental life (subjectivity) and a neurophysiological explanation of man's organic body, without need of any recourse to psychical or physical causal nexuses between mind and body.

I agree with Jonas that psychosomatic pathology presents legitimate phenomena for investigation, though in my view it is not made so because psychogenic factors *cause* physiopathic symptoms. Quite simply, all infirmity may be conceived as 'psychosomatic' or 'somatopsychic,' those terms not to be used to indicate another subspecialty of medicine, however. Once we refuse (with Husserl) to construe psyche and soma as substances having causal efficacy 'running both ways,' psychosomatic and somatopsychic illnesses become subsets or extreme limiting categories of all infirmity. Thus we can correctly say of some cases that (1) certain physical manifestations of a patient's organic body may come into focus when the etiology suggests their psychical significance; for some others (2) certain psychical manifestations of the patient's mental life may become focal when the etiology suggests their physical significance; and of still many others (3) both the psychical and physical components of the ailment may be so intimately conjoined that the physician will (for good diagnostic and clinical reasons) refuse to subsume the infirmity to the standard dichotomous taxonomy – the functional and organic. Thus all neurophysiological processes have psychical significance and all psychological processes have a neurophysiological foundation. Hence, whenever abnormal psychic manifestations are present, the physician may approach a patient as one with mental illness, which has its root in the psyche in so far as the latter uses bodily organs; it is also rooted in the body in so far as the latter is the organ of the psyche. All illness produces some disturbance within the complex set of relations – psyche and soma.

IV. NEURO-MEDICINE AND THE PSYCHE-SOMA DISTINCTION

The problem of the relationship between body and mind is as old as medicine itself. Since Aristotle, medicine has struggled to find an integrative formula, an approach which would offer both a methodological unity and an empirical coherence for clinical practice. Historically, physicians interested in investigating the psyche faced the choice of either borrowing heavily from the philosopher or remaining within the narrower confines of anatomy and rather rudimentary neurophysiology.

By 1952, in his *Rede Lecture* at Cambridge, Russell Brain remarked that "medicine has reached a point at which it cannot escape certain philosophical questions, for it would seem either that the physical and psychological explanations of mental states cannot both be right, or that, if they are, there must be some logical explanation of their relationship to one another" ([1], p. 20). The import of Hans Jonas' presentation in the context of a Symposium on the Philosophical Dimensions of the Neuro-Medical Sciences is revealed in his thought experiment, which has as its central concern the relation that obtains between our minds and brains. It is encouraging to note that Jonas is joined in his effort by pharmacologists and biomedical scientists, some of whom represent *materia medica*, as well as clinicians who articulate the *praxis medica* as practicing neurologists and neurosurgeons.

In *Facing Reality: Philosophical Adventures by a Brain Scientist*, John C. Eccles reminds us that it has been fashionable of late for philosophers "to discredit or even deride all problems purporting to derive from the concept of mind or of consciousness" ([5], p. 64). Eccles takes it as "very encouraging to neurophysiologists" that a "counter-attack" is on the horizon led by neuromedical and neuroclinical scientists. "For many of us," he says, "despite the philosophical criticism, have continued to wrestle with the problem of brain and mind, and have come to regard it as the most difficult and fundamental problem concerning man" ([5], p. 64). In this sense Eccles and Jonas are in agreement as to aims. But it matters a good deal whether the standard formulation of the psychophysical problem is really a pseudo-problem, i.e., the initial formulation of the "problem" has rested on mistaken assumptions or (should I say) on mistaken conceptions of 'mind' and 'body.' The important point is that the implications of one's conception of the psychophysical relation are far-reaching and of great significance, not only with respect to theoretical considerations in medicine, but in medical practice as well; for depending on one's conception of what we are, whether mechanisms or persons, essentially embodied minds or essentially physical bodies, mere complex material conglomerates in Nature or emerging cultural entities, we can count on the fact that a response will always be expressed by those who participate in *praxis medica* as long as there is medical practice. One illustration of this is revealed in the domain of neurological medicine: Consider patients with stroke or epilepsy. It is possible to follow the view of the epiphenomenalist, taking consciousness, subjectivity, or mentation and life function as the mere penumbra of the anatomical state of affairs, and to consider in a mechanistic way the brain and CNS as the total clinical phenomenon. This is to take a single-track approach[3] to the insult or aberration and to ask only 'Where is it?', 'What is it?' and, having answered these, to intervene with aggressive drug therapy. For such a view no notion of subjectivity seems to be required – it is, in fact, rejected as "unreal," an "epiphenomenon." Neurophysiologists and neurologists may, of course, *hope* to provide a complete account of what goes on during any mental process, feeling secure in ignoring subjectivity and mental events altogether. But how different is the neurologist who undertakes to answer another question: 'What is the injury doing to the patient, his entire life functioning, his general life situation, including his mental life (which is not a mere epiphenomenon)?'

There was a time prior to the development of specific agents (in which the entire thrust was to discover particular agents to eliminate pathogenic symptoms) when the major part of neurological diseases were subsumed under the category of chronic and even incurable diseases and when the physician included into medical assistance those incurable or hopeless cases such that medical assistance was extended from the cure of the sick to the care for the infirm. With the discovery of particular and specific etiological agents whose prescription was isomorphically related to the particular insult or neural aberration, the neurologist was content with his treatment modalities and often abandoned the intimately conjoined unity we call the patient, ignoring the long-term management of his or her infirmity. Thus instead of prescribing and administering the armamentarium of specific agents *in combination with* the functional totality constituting the patient's full unity of psyche and soma, the patient was treated one-sidedly, *treated* somatically. The result was therapeutic nihilism,[4] a shift in medical praxis, not just theory. Neurology and neurosurgery (I suggest) have to relearn the maxim: *cessante causa non cessat morbus* (the elimination of the cause does not necessarily entail the end of the disease); they cannot hide behind the skirt of a crude definition of neurology's subject matter, which has it that it is only concerned with functions of the CNS as affected by focal or localized lesions and/or general systemic disorders. Such a restrictive definition only pretends, however, to justify the purchase and employment of equipment to perform axial transversal computerized tomography which, paradoxically, will reveal in greater detail answers to the questions 'Where is it?', 'What is it?' I say *paradoxically* since the employment of computerized tomography may well spell the end of any remaining difficulties in answering these questions of locale and type of insult. Could it be that the advent of this precise computerized radiology will compel neurologists (for the wrong reason) to pursue the more difficult question of the entire life function and management of their patients? Although the methodological approach of John Hughlings Jackson, who worked from the assumption of a strictly somatic neurological science, was, in his historical context, quite appropriate and even necessary,[5] today the demands of the neuroclinic make themselves manifest. Thus neurotherapy in whatever form must aim at restitution of the infirm patient, even if *restitutio ad integrum* is not fully achievable. Surely stroke victims will never quite return to a state indistinguishable from what they were before the insult. Hence, I am not suggesting that clinicians engender false hopes in their patients by pretending they can guarantee a higher frequency of cure, especially if they apply strictly clinical criteria. But the history of neurology suggests that the view we have of ourselves as either essentially anatomical, physical bodies *or* as essentially psychosomatic unities whose subjectivity engenders social and cultural life, carries with it implications for clinical praxis. The chapter in the history of neurology which will one day reveal today's *praxis medica* may be none too flattering. In fact, an archivist may well come across a medical record on which has been preserved the notes of a clinician who, during grand rounds, said of his alcoholic patient, who was responding to treatment, that the "mental brain is okay again." We should admit, in all fairness, that such language is no less muddled than that of the bruised alcoholic who, after additional misfortunes in the bar, added to his medical history by remarking: "I'm this way because they threw me out bodily."

V. APOLOGIA

If any of the preceding has generated pseudo questions for Professor Jonas, I ask his pardon; it is quite natural, though surely improper, for commentators to resist the annoying superiority of genius.

University of Connecticut Health Center,
Farmington, Connecticut

NOTES

[1] Husserl: 1952, *Ideen zu einer reinen Phänomenologie und phänomenologischen Philosophie*, Book II, Martinus Nijhoff, The Hague, esp. secs. 35-42, 62-63. Also see Husserl: 1958, *Ideas* (transl. by N.R. Boyce Gibson), MacMillan Co., New York, p. 419.

[2] For an account of lived-bodily categories see my 1975, 'Lived-Body as Catalytic Agent: Reaction at the Interface of Philosophy and Medicine', in H.T. Engelhardt, Jr. and S.F. Spicker (eds.), *Evaluation and Explanation in the Biomedical Sciences*, Vol. I, *Philosophy and Medicine*, D. Reidel, Dordrecht-Holland/Boston-U.S.A., pp. 181-204.

[3] A point I owe to Ian Lawson, M.D.

[4] Riese, W.: 1959, *A History of Neurology*, MD Publication, New York, esp. ch. IX, 'The History of Therapy in Neurology', pp. 174-180. Also see Walshe, F., Sir: November 24, 1956, 'The Nature and Dimensions of Nosography in Modern Medicine', *The Lancet*, pp. 1059-1063. In fairness to neurologists it should be noted that "A recent medical manpower analysis indicates that neurologists comprise less than 1 percent of all physicians...". Thus, the existing shortage of neurologists contributes significantly to the fact that "a neurologist sees approximately 860 new patients per year, devoting an average of 1.8 hours per patient." Given these figures, is it any wonder that neurologists have not been able to undertake long-term rehabilitation programs with their patients? This situation too must be remedied. Cf. Yahr, M.D.: June 1975, 'Summary Report of the Joint Commission on Neurology,' *Neurology* **25**, 497-501.

[5] Cf. Engelhardt, H.T.: 1975, 'John Hughlings Jackson and the Mind-Body Relation', *Bulletin of the History of Medicine* **49**, 137-151.

BIBLIOGRAPHY

1. Brain, R.: 1952, 'The Contribution of Medicine to our Idea of the Mind', *Rede Lecture*, Cambridge Univ. Press, Cambridge.
2. Descartes, R.: 1967, *The Passions of the Soul*, arts. XXXII, XXXV, XLIII (transl. by E.S. Haldane and G.R.T. Ross), *The Philosophical Works of Descartes*, Vol. I, Cambridge Univ. Press, Cambridge, pp. 346-347.
3. Descartes, R.: 1969, Letter 'Princess Elisabeth to Descartes (10-20 June, 1643)', in E. Anscombe and P.T. Geach (eds.), *Descartes: Philosophical Writings*, Thomas Nelson and Sons, London.
4. Eccles, J.C. (ed.): 1966, *Brain and Conscious Experience*, Springer-Verlag, Berlin-Heidelberg-New York.
5. Eccles, J.C.: 1970, *Facing Reality: Philosophical Adventures by a Brain Scientist*, Springer-Verlag, New York-Heidelberg-Berlin.
6. Engelhardt, H.T., Jr.: 1973, *Mind-Body: A Categorial Relation*, Martinus Nijhoff, Hague.
7. Jonas, H.: April 1965, 'Spinoza and the Theory of Organism', *Journal of the History of Philosophy* 3, 43-57; reprinted in S.F. Spicker (ed.): 1970, *The Philosophy of the Body*, Quadrangle Books, Chicago, Ill., pp. 50-69.
8. Jonas, H.: 1974, *From Ancient Creed to Technological Man*, Prentice-Hall, New Jersey.
9. Jonas, H.: 'On the Power or Impotence of Subjectivity', in this volume pp. 143-61.

SECTION V

ALTERED AFFECTIVE RESPONSES TO PAIN

GEORGE PITCHER

PAIN AND UNPLEASANTNESS

What is pain? In one sense of the term 'know,' we all know what pain is; we can talk about it intelligibly, we can recognize genuine instances of it in others and in ourselves, we can inflict it, almost anyone can relieve simple cases of it, and medical experts can relieve many more kinds. But it is not easy to put this knowledge into words, that is, to *say* what pain is. We in philosophy are familiar with this puzzling kind of phenomenon; it has always been our daily fare. Consider, for example, what Augustine says about time:

> For what is time? Who can readily and briefly explain this? Who can even in thought comprehend it, so as to utter a word about it? But what in discourse do we mention more familiarly and knowingly, than time? And, we understand, when we speak of it; we understand also, when we hear it spoken of by another. What then is time? If no one asks me, I know: if I wish to explain it to one that asketh, I know not... ([2], p. 253).

I doubt that pain is as difficult as time, but it is hard enough. In any case, let us persist with our question: What is pain?

Well, it is a certain kind of sensation. But *what* kind? One is tempted to answer that it is a sensation that has its own characteristic felt quality, *Q*. This is undoubtedly true, but it tells us nothing. Sensations are things that are felt; indeed, they exist if and only if they are felt. This used to be said of desires and thoughts as well, until Freud convinced us, or some of us, that we have to acknowledge the existence of unconscious desires and thoughts: but so far there has been no need to countenance unconscious (i.e., unfelt) pains and itches.[1] So sensations necessarily feel *somehow or other* – i.e., they have a felt quality – and different kinds of sensations feel different. This is a boring truth. But to declare that pain is the kind of sensation that has its own characteristic felt quality, *Q*, is just to repeat that boring truth. It is to say no more than that pains are the kind of sensation that feel the way pains do.[2]

We need to find a more informative way to characterize pain. We might try this: pain is what one feels when some of his body tissues are either injured (or damaged) or in danger of becoming injured (or damaged). Here we have an attempt to characterize pain by reference to its alleged cause. I

S. F. Spicker and H. T. Engelhardt, Jr. (eds.), Philosophical Dimensions of the Neuro-Medical Sciences, 181–196.

believe it is true that pains *are* standardly caused by body tissues that are either in one way or another damaged or else are in danger of becoming so. This is a fairly straightforward empirical claim that seems to have been widely, though not universally, accepted for some time (e.g., [9], [11]). I have even defended in print the view that to feel a pain is, in the normal case, to perceive the damaged, or near-damaged, state of a part of one's body [7]. If this is right, as I still think it is, then the feeling of pains, like the seeing of colors or the hearing of sounds, is a form of sense perception. The feeling of pains is a form of sense perception, however, only in the normal or standard case, because there are well-known examples of pains that are not caused by any tissue damage (or near-damage) in the part of the body where the pain is felt (as in phantom limb pain and referred pain) and examples of pains that, as far as we know, are caused by no tissue damage (or near-damage) at all (as in hysterical pain and pain induced by hypnotic suggestion).

What, now, about the unpleasantness, or worse, of pains? Is this a feature of absolutely all pains or only of the usual, or standard, cases? Here we encounter one of the truly baffling questions about pain – the kind of question that sends the philosophical mind into action. It will in fact be my main concern in this paper. On the one hand, what could be more obvious than that pains are, at best, unpleasant and, at worst, absolutely unbearable? Just imagine a pain – a toothache, say; now how could *that* sensation be anything but unpleasant? Being unpleasant, or worse, seems to be part of the very essence of pain. And yet, on the other hand, there seem to be many different cases of pains that are not unpleasant and, indeed, some that are even enjoyed. Here are some well-known examples:

1. The pains of the masochist.
2. The pains of prefrontal lobotomy patients who claim not to mind them.
3. The pains of Indian fakirs who can lie on beds of nails or walk barefoot over hot coals without minding the pain.
4. The pains of certain patients to whom an analgesic, such as morphine, has been administered. Such patients commonly report that their pain is as great as ever, but now they no longer mind it.
5. The pains of patients suffering from asymbolia for pain. Here again, the patients report feeling pain, but they do not seem to mind it.

In none of these cases is it absolutely certain that the person both (a) does feel a pain, and (b) does not mind or dislike it. For example, consider asymbolia for pain; it could plausibly be argued that these patients do dislike their pain but are simply unable to "organize their behavioral responses for withdrawal or some other appropriate action," as Sternbach puts it ([10], p. 97). Again, if we accept the Melzack-Wall gate control theory of pain – and I assume that it is the best physiological theory of pain that has been proposed so far – then one could argue that fakirs have somehow learned to put themselves into a state in which, among other things, their pain gates are closed, so that they feel no pain at all. (In that case, we could regard them as being self-taught pioneers in the modern practice of biofeedback conditioning!)

I cannot discuss all of these different cases in the detail that they deserve. Instead, I shall concentrate on the first – the pains of the masochist – to see if I can discover whether they really are, in fact, pains that are not unpleasant (or worse).

Note first that it is just possible that the pain gate of the masochist (when he is, as it were, practising his art) is entirely closed, so that he feels no pain at all, in which case the masochist would not constitute a counterexample to the thesis that all pains are unpleasant, or worse.[3] The question of whether the masochist's gate is closed or not is, of course, an empirical question. Let us grant that one relevant fact is that masochists claim that they do feel pain. This counts against the hypothesis that their pain gates are closed. But suppose that it were possible to implant electrodes in the brains of masochists, and suppose further that in countless experiments of unquestionable reliability, it was found that no signals from the central transmission (*T*) cells of the gate control system (i.e., roughly, no pain signals) were reaching the brains of the masochist subjects. The most reasonable thing to think in these circumstances might well be that the masochist feels no pain, despite his protestations to the contrary.

I once suggested, in a paper [6], that if the Melzack-Wall theory of pain is correct, then the pain gates of masochists must be closed. S. Noren and A. Davis, in criticizing this claim, argue that there are two important conceptual difficulties that make it unacceptable.

> First, if his gate is closed then the masochist feels no pain. But what sense is there, then, in calling him a masochist? There would be no such persons as masochists, or anyone else fitting the description of a person who enjoys inflicting pain on himself. [5].

But this is not a real difficulty. Noren and Davis suppose that there could then be no such thing as a masochist because they assume that a masochist *has* to be thought of as one who enjoys having pain inflicted on him. But if we were faced with the imagined strong evidence described above, the reasonable thing to do would be to cast aside our earlier conception of a masochist and think of him instead as one who enjoys being humiliated or perhaps as one who has a deep need to be debased and mastered and as one who can achieve sexual gratification only if he is so debased.

The second difficulty that Noren and Davis claim to find in the supposition that the pain gate of the masochist is closed is that we could then make no sense of "the masochist's claim that he *in fact* feels pain when being whipped, and likes it" (p. 122). They contend further that "if... the masochist claimed to be in pain, we would side with him, though if [the Melzack-Wall theory] were true and applicable we might have reason to believe his pain is not as severe as a person with normal desires would have in that context" (p. 122). But if we had our imagined strong evidence, we would not be wise to side with the masochist, and we would have to find some explanation for his apparently sincere pain avowals other than that he acually does feel pain. What such explanations might be is not at all difficult to imagine. One obvious candidate would be this: the masochist's cries of pain are part of his fantasy, in which he imagines that he is being punished for some fault and that he is suffering great and deserved torment. He has a deep need to think that *pain* is being inflicted on him for his supposed guilt, and so he imagines that he really does feel it, whereas in fact he does not. We might even add that it is part of the masochist's fantasy that his 'master' should want him, the masochist, to suffer great pain as a result of the master's actions towards him; this would give the masochist an additional motive for thinking, and for saying, that he is in pain, though in fact he is not.

So I think it is conceivable, and in every way possible, as I suggested in my original paper, that the masochist feels no pain. But in the absence of the imagined strong evidence to support that hypothesis, I have to confess that it now seems to me to be highly unlikely that it is true; for although I have never witnessed a masochistic orgy (not even in the films) – i.e., I have not done my field work – I believe, from my reading of the literature, that the masochist typically winces, groans, and cries out when the lighted cigarette ends, whips, or whatever, are applied to his flesh, and the most plausible explanation of this behavior, in light of all we know of the masochist, would seem to be that he is feeling genuine pain.

But now we must face the question: Are the pains of the masochist unpleasant or not? I want to approach this question by trying to answer first the preliminary question: Does the masochist dislike his pains or not? It seems to me that there are two plausible hypotheses to be considered:

(a) that a masochist likes, indeed positively enjoys, his special pains, and

(b) that the masochist dislikes his pains.

There is the further possibility that a masochist neither likes nor dislikes his special pains, but there seems to be no evidence supporting this hypothesis, so I shall not consider it.

We naturally assume that (a) and (b) cannot both be right, and that if we choose one, we cannot accept the other. Let us proceed on this assumption for the time being.

There are forces pulling us in the direction of hypothesis (a) and others pulling us in the direction of (b). In favor of (a) are the facts that a masochist seeks out situations in which his special pains will be inflicted on him, and that he would undoubtedly be sorry if his partner prematurely stopped torturing him. Furthermore, many masochists say that they enjoy their pains, (see [3]) and if these reports are to be trusted, they seem to confirm (a) at once. And it is certainly true that we do generally trust people as the best judges of what they do and do not enjoy, so that special explanations must be given whenever we go against this principle. Finally, it is probably true that something that is originally disliked, when it is habitually associated with something else that is liked, can become itself liked – and this principle might well apply to the masochist's pain.

The case for (b) seems even stronger. First of all, there is strong behavioral evidence that tends to confirm (b): the way a masochist behaves while he is being whipped, burned, or whatever, seems to show that he is experiencing sensations that he intensely dislikes. It might be objected that this behavior shows only that he is experiencing *pains*, and that it cannot, without begging the question, be used twice over to show that he also dislikes these pains. But there are aspects of the masochist's behavior that support the additional hypothesis that he dislikes his pains, as the rest of us would. For what is being claimed when it is claimed that a person dislikes a sensation? As far as I can see, what is claimed is just that the person has a spontaneous, or unreasoned, desire that the sensation stop. He may have, or think he has, good reasons for wanting it to continue, so that *all things considered*, he does not want it to stop, but as long as he has a spontaneous desire that it stop, he dislikes the sensation. For example: when the dentist is drilling without having administered an anesthetic, we usually feel a sensation that we dislike having – in fact, a pain; we have a spontaneous inclination to want it to stop, but all things considered, of course, we do not want it to stop, for we want the tooth to be fixed. But the masochist does, I believe, exhibit behavior that indicates that he has a spontaneous desire that his pains should stop: he contorts his body to avoid the lash or the burning cigarette, he begs his 'master' to stop the 'discipline,' he cries out that it is terrible, that he cannot stand it any longer, and so on. So we have good reason to think that the masochist dislikes his special pains.

Here, then, we apparently have a genuine puzzle – for the masochist seems both to like or even positively to enjoy his pains and at the same time to dislike them. It is true that some of the evidence adduced in favor of hypothesis (a) – viz., that the masochist enjoys his pains – is perfectly compatible with hypothesis (b) – viz., that the masochist dislikes his pains. Thus, consider the fact that a masochist seeks out certain special situations in which pain will be inflicted and would surely be disappointed if his tormentor were prematurely to stop beating or burning him. This fact is not evidence against the view that the masochist dislikes his pains, for we have an entirely plausible hypothesis, one that is well entrenched in psychological theory, to explain this aspect of the masochist's behavior – and it is a hypothesis, moreover, that virtually requires that the masochist dislike his pain. I mean the hypothesis that the masochist has a deep need to be *punished* – i.e., to suffer something that he dislikes – for some imagined wrongdoing, and that unless he really is punished, he is unable to achieve sexual gratification. On this hypothesis, his seeking out of his special pains and his being disappointed if his tormentor were to stop inflicting them on him, may not show that he enjoys those pains, but rather that the pains satisfy a deep need for punishment and lead to sexual fulfillment.

The other main fact that we thought supported hypothesis (a) is that many masochists report that they enjoy their pains. I can think of no plausible reason for doubting these reports – nor does there seem to be any plausible way of reconciling them with hypothesis (b), as long as we continue to assume that (a) and (b) are inconsistent with one another. What, then, are we to say about this baffling situation?

I suggest that we must take another look at the assumption we have been making that hypotheses (a) and (b) are incompatible with one another. And when we do, we shall find, I suggest, that they are not, that there really is nothing self-contradictory in the claim that a person can enjoy something that he dislikes. Indeed, I think the masochist's claim that he enjoys his special pains is entirely consistent with the supposition that he dislikes them and, hence, with hypothesis (b).

To this it might be objected that I have claimed that the masochist, in so far as he dislikes his pains, has a spontaneous desire that they stop, whereas surely to *enjoy* something entails at least that the person has a spontaneous desire that it continue, and therefore that it *not* stop. But this is no objection, for it is perfectly possible for a person to have two contradictory spontaneous desires concerning something. Indeed, I think that it would be a very apt characterization of the pathetic condition of the masochist to say that although he has a spontaneous desire that his pains should stop, still, in his sexual frenzy and for causes with which we are familiar, he has an even stronger impulse to want them to continue (i.e., he positively enjoys them). I think it is safe to say that many of us, in our sexual activity, experience precisely this apparently contradictory attitude toward pains that are less intense than those of the masochist. It would seem, then, that the natural assumption with which we began our discussion of hypotheses (a) and (b) is false – I mean the assumption that (a) and (b) are incompatible with one another. (a) and (b) may both be true; indeed, I think that the most reasonable conclusion to draw from the available evidence is that (a) and (b) *are* both true.

My proposal, then, is to reject the assumption we tend naturally to make, that hypotheses (a) and (b) must be incompatible with one another. The most plausible alternative to this proposal, I think, is to cling to the original assumption and explain away the main support of (a) – i.e., the masochists' reports that they enjoy their pains. One could hold that the masochist is deceived when he thinks he enjoys his special pains. He takes his pains to be the object of his enjoyment, whereas in reality it is something else that he enjoys – the promise of sexual gratification, being mastered or humiliated, or whatever ([8], [3]). The truth of the matter, according to this alternative proposal, then, is that the masochist does not enjoy his pains at all. He merely endures them because he thinks he deserves them and because they promise some expected pleasure (that of sexual fulfillment). On this way of looking at the masochist, he is likened to the ordinary person who goes to the dentist to have a tooth extracted; the dentist's patient does not enjoy the pain involved, but he puts up with it for the sake of a healthy mouth. As Reik puts it:

He [the masochist] does not enjoy pain, but what is bought with pain! He does not strive for discomfort, but for lust that must be paid for with discomfort.[4]

But I do not think that this view can be right. I shall try to show that the status of pain in the masochist's case is quite different from its status in the case of the dentist's patient. For the dentist's patient, pain is not even a means to a desired end; it is a mere accompaniment of a means to a desired end. (The dentist's drilling is the means; the pain is a mere accompaniment of that). The dentist's patient merely puts up with the pain for the sake of a desired end; for to say this is to say that (i) he dislikes the pain, and (ii) if his situation were to remain exactly the same, but the pain were to be eliminated, he would greatly prefer it – and both of these things are obviously true of the person in the dentist's chair.

But what about the masochist? Is his pain a mere accompaniment of a means to a desired end? And does he merely put up with the pain in order to achieve that desired end? Well, the pain *is* undoubtedly an accompaniment of the whipping or the burning, and these latter are certainly means to an end desired by the masochist (viz., orgasm). But does the masochist merely put up with the pain in order to achieve orgasm? Well, if he does, then it must be the case that (i) he dislikes his pain, and (ii) if his situation were to remain exactly the same, but the pain were to be eliminated, he would greatly prefer it. We have already acknowledged the truth of (i); but it seems unlikely that (ii) is true of the masochist. I imagine that if his situation were to remain exactly the same, except that the pain were to be eliminated, he would *not* prefer it; on the contrary, all the excitement would be gone. I suppose that, in principle, one might discover whether I am right about this by conducting some clever experimental tests. If I am wrong, then we might well be entitled to say that the masochist does not enjoy his special pains but simply puts up with them. Masochism would then not be as philosophically interesting nor as conceptually difficult a phenomenon as I take it to be. But the fact that masochists say that they enjoy their pains leads me to think that they would almost certainly *not* prefer to be burned, whipped, and so on, by their sexual partners, without pain. If this is right, then the masochist cannot be likened to the person who endures pain in the dentist's chair or at the hands of the surgeon.

But it has not yet been shown that the masochist *enjoys* his special pains. We can say now that the masochist's pain is not a mere accompaniment of a means to a desired end, as pain is for the dentist's patient. But perhaps the masochist's pain is itself a *means* to a desired end, and perhaps the masochist merely puts up with it because he is unable to achieve that end without going through the pain. Well, I think his pain certainly *is* a means, or at any rate, part of a means, to an end desired by the masochist – namely, orgasm.[5] Now, if orgasm were his *sole* desired end, then a plausible case could probably be made out for the thesis that the masochist does *not* enjoy his pains, and that he merely puts up with them because he unfortunately cannot achieve orgasm without them. But orgasm is of course *not* the masochist's only desired end; there is also the process of getting there, which is for the masochist, as it is, or anyway should be, in non-masochistic sex as well, highly exciting, and is therefore desired in part for itself. For most masochists, the excitement comes from the acting out of a fantasy – e.g., that he is being humiliated and debased by his partner, or that he is the helpless slave of his 'master.' Something like that. But it is surely wrong to suppose that the masochist merely endures his pains so as to fulfill this fantasy; for that would imply that he would prefer the whole scene if he could fulfill the fantasy without going through the pain – and this seems clearly false. The masochist could easily satisfy the fantasy that he is a slave, mere putty in the hands of his 'master,' without enduring any pain; for instance, his 'master' could force him to run around the room yelling "oink," or make him write "I am a bad boy" 100 times on the blackboard. That just might turn somebody on, but your typical masochist – the one I am concerned with, at least – is not going to get much of a thrill out of that kind of pale performance. No, for a fully satisfying sexual experience, the fantasy that he is a helpless slave must be fulfilled by the infliction of *pain*. Otherwise, the thrill is gone, or anyway seriously diminished. So the masochist does not merely endure his pains; he is thrilled and excited by the whole scene, a scene that includes pain, caused in a certain way, as an essential part. In this sense, then, he positively enjoys his pains, just as he says he does.

We can perhaps understand the masochist a bit better if we think of our own more prosaic experiences on a roller coaster, where we positively enjoy feelings of fear. Normally, we intensely dislike such feelings, but not on a roller coaster. Indeed, they are what we are primarily paying to get when we buy our ticket. It is no use objecting that we do not really enjoy the feelings of fear but merely put up with them in order to get the exhilaration and thrills, for to say that is to say that if one were to imagine that everything were left the same except that the feelings of fear were to be eliminated, we would prefer that imaginary roller coaster ride to the actual one. This is wrong, because if you remove the feelings of fear, you remove the peculiar thrill; the special excitement of the roller coaster ride is the exhilaration of feeling afraid. (A vague feeling of thrill, with nothing one is thrilled about or by, even if such a thing were conceptually possible, would not be at all the same thing.) So although we dislike feelings of fear in most situations, in *this* setting (i.e., on the roller coaster), they exhilarate and thrill us – that is, to put it in the widest, and blandest, relevant category, we *enjoy* them. Just so with the masochist, I suggest: although he, like everyone else, merely dislikes pains in most cases, in his special sexual scene, he is also thrilled and excited by them – to put it mildly, he enjoys them.

I am sure that there will be lingering doubts about this in some readers' minds. Their misgivings might be expressed as follows: "To enjoy something, one has to like it for its own sake and not just for the sake of something else to which it contributes or which it helps to produce. It seems, however, that the masochist does not like his special pains for their own sake, but just because they contribute to his state of sexual excitement. How, then, can you say that the masochist *enjoys* his pains?" To this, I reply that the notion of enjoying or liking a pain for its own sake is terribly unclear; I do not know what that expression is supposed to mean. Is it supposed to mean that the masochist should like his pain even if all its usual accompaniments were to be stripped away – i.e., if there were no sexual partner inflicting the pain, no humiliation, no orgasm, but just the pain, all by itself? Surely this is asking too much: if so much were required in order to enjoy a thing, it is unlikely that anyone could be said to enjoy anything. My claim is simply that *in a special setting*, the masochist likes his pains; if the pain were to be eliminated from that setting, he would not value the experience at all or anyway not nearly so much. As far as I can see, this is all that is required for the truth of the proposition that the masochist enjoys his pains. Similarly, if someone likes olives in a paella, then he likes paellas, and he considers paellas without them to be not as good. Does this mean that he likes the taste of olives for its own sake? Again, I do not know what that means here. But surely *in a paella* he likes them. Just so does a masochist, in my view, like (or enjoy) his special pains: *in a certain setting*, he likes (enjoys) them.

We cannot, then, tidy up the masochist; that is, we cannot explain away the facts that show that he enjoys his special pains, in order to maintain that he *simply* dislikes his pains, thus putting him in a familiar category of cases. Nor, on the other hand, can we explain away the facts that indicate that he dislikes his special pains, in order to maintain that he *simply* enjoys his pains. One or the other of these moves may seem to be necessary to avoid what looks like the contradiction of asserting that a masochist both likes (enjoys) and dislikes his special pains. Certainly many philosophers would urge that such conflicting attitudes are logically impossible. Thus D.M. Armstrong writes:

> Consider the masochist... It is not the case that 'his pain is his pleasure'. Taken literally, this implies that he has a favourable attitude to the thing he has an unfavourable attitude to, which is contradictory. ([1], p. 91)

But there is no contradiction here. There might be a contradiction if one tried to hold that a masochist likes his pain for a certain reason and dislikes it for the very same reason; but there is no contradiction in saying, as I do, that the masochist likes his pain for a certain reason and dislikes that very same pain for a different reason.

The belief that there is a contradiction involved in claiming that the masochist both likes and dislikes his special pains inevitably leads to some ignoring, or artificial explaining away, of evidence, and, therefore, to a one-sided view of the masochist and a false view of pain itself. Thus on the next page after the quotation above, Armstrong says: "... It is of the *essence* of pain that we wish it to stop" (p. 92). Now, as far as one can determine just by considering the masochist alone, this statement could be true; for, of course, the masochist *does* wish his pain to stop. But once we recognize that his wishing it to stop is not incompatible with his wanting it to continue, we shall be very suspicious of Armstrong's thesis about the essence of pain – for we shall be able to see the possibility of there being people who do not dislike their pains, but who *merely* enjoy them, and the possibility of there being people who neither like nor dislike their pains, that is, who "don't mind them." Prefrontal lobotomy patients, evidently, are actual examples of the latter sort of person. But thinkers like Armstrong who suppose that it is contradictory to assert that a person both likes and dislikes his pain, and who settle for the view that "it is of the *essence* of pain that we wish it to stop," are going to be embarrassed by all those cases in which people seem *not* to dislike their pains. For example: given Armstrong's thesis about the essence of pain, he ought to deny that lobotomy patients feel pain; realizing that this is too implausible, however, he says:

In such an unusual situation, a linguistic decision has to be taken as to whether they can be said to be feeling 'pain' or not ([1], p. 108).

But this, too, is implausible, for there is no good reason for denying that lobotomy patients feel real pain, in the perfectly ordinary sense of that term.

We have now answered what I called the preliminary question: "Does the masochist dislike his special pains or not?"; our answer has been that he both dislikes and likes (enjoys) them. What consequences does this answer have for the main question: "Are the pains of the masochist unpleasant or not?" Well, if the masochist dislikes his pains, this is surely because he finds them unpleasant. There is, I think, a temptation to suppose that if a person finds a sensation unpleasant, then, necessarily, it *is* unpleasant. The idea seems to be that a sensation is a private, or mental, object, about which the person whose sensation it is cannot be mistaken; so if he takes it to be unpleasant, how could it fail to *be* unpleasant? But then we seem to be in logical trouble again, for we have argued that the masochist also *enjoys* his pains. If he enjoys them, surely this is because he finds them enjoyable or pleasurable; but then, by parity of reasoning, must not they *be* enjoyable or pleasurable? But how can something be at the same time both unpleasant and enjoyable (pleasurable)?

The source of the trouble here might, of course, be the thesis that the masochist both likes and dislikes his special pains; but I do not think so. I think it lies, rather, with the assumption that if a person finds a sensation unpleasant, then necessarily it *is* unpleasant. It is certainly true that in some sense or other – and it is not at all clear precisely what the sense is (a huge philosophical issue!) – each of us has some kind of special privileged access to his own sensations. But from this it by no means follows that for *any* property whatever of a sensation, if a person takes a sensation of his to have that property, then it necessarily does have it. There are ever so many properties of sensations for which this principle is plainly false – e.g., the property of being caused by the firing of such-and-such nerve fibres. I want to argue, now, that the property of being unpleasant belongs to the same class.

What sort of property could unpleasantness be? It is extremely implausible to suppose that it might be an inherent or intrinsic property of the things we call 'unpleasant,' as the shape of a table, or the pitch of a sound, are intrinsic properties of those things. There is a vast range of widely disparate things that can be called 'unpleasant' – smells, color combinations, sounds, sensations, persons, incidents, encounters, meals, sights, locations, rides, and so on. It just seems extraordinarily unlikely that there is any intrinsic property or any cluster of such properties that unpleasant individuals of these wildly different kinds all have and that is picked out by the term 'unpleasant.' What seems far more likely is that when we speak of something's being unpleasant, although we usually have in mind some features of the thing in virtue of which it *is* unpleasant (e.g., the dinner was unpleasant because the host got drunk and had words with the butler), what we mean is that the thing, because of those features, elicits some kind of adverse reaction – we might as well call it 'disliking' – from some person or group of persons. To call something 'unpleasant' is not to make a totally subjective judgment; e.g., if I say that the party was unpleasant, I am not claiming merely that *I* disliked it. Others who were there might dispute my claim about the unpleasantness of the party, and the issue between us would seem to be a genuine one on which only one of us could be right. This is supported by the fact that there are such expressions in the language as "Well, *I* think it was unpleasant" and "He found the interview very unpleasant," which imply a distinction between 'appearance and reality' – that is, between merely *thinking* that something is unpleasant (or merely *finding* it unpleasant) and its really *being* unpleasant. The best hypothesis, I suggest, to account for all the relevant facts is that "x is unpleasant" means something like this: x is the sort of thing that most (or all normal) people would dislike, or x has a propensity to be disliked by most (or all normal) people (cf. [12]).

If this hypothesis is right, then our conclusion that the masochist both likes and dislikes his special pains does not saddle us with a contradiction. The masochist has the normal reaction to pain: he dislikes it. But because of his psychological make-up, he also has a quite different reaction to his pains, for he positively enjoys them. This latter fact does not mean that his pains are not unpleasant, however; it means only that the masochist happens to enjoy something that is unpleasant – i.e., something that most people would dislike. Our analysis of unpleasantness makes this perfectly intelligible. (Similarly, the smell of rotten eggs would be an unpleasant smell even if there happened to be people who did not mind it or who even delighted in it.)

I have indicated one kind of conceptual trouble that one can fall into if he rejects my hypothesis about the meaning of 'unpleasant' and holds instead that whenever anyone dislikes one of his sensations, that sensation *is*, necessarily, unpleasant, and, conversely, whenever anyone does *not* dislike one of his sensations, that sensation fails, necessarily, to be unpleasant. This second view can lead to other errors as well. Consider what it implies about lobotomy patients, for example. These patients say that they feel pain but do not mind it – i.e., do not particularly dislike it. Someone who holds the view I am criticizing will have to say that the lobotomy patients are confused, or anyway wrong, since what they feel cannot be pain. The reasoning goes like this: "These patients do not dislike the sensation they call 'pain'; therefore, the sensation cannot be unpleasant. But all pains are unpleasant. So what they feel cannot be pain." But there is no good reason for thinking that lobotomy patients are wrong when they say they feel pain and every reason to regard their reports as accurate. Once it is seen that their not minding their pains does not at all entail that those pains fail to be unpleasant, there is no longer any need to embrace the implausible view that lobotomy patients are mistaken when they report pain.

Here is another case where trouble is caused by treating the unpleasantness of a pain as if it were the same thing as its being disliked by the person whose pain it is. Melzack and Casey [4] hold that pain consists of a sensory dimension and an affective (or a motivational-affective) dimension;[6] that is, to have a pain, a person must have a certain kind of sensation, *and* he must dislike it. ("... The motivational-affective processes... are an integral part of the total pain experience." [p. 425]) As far as I can tell, the only reason Melzack and Casey could have for making the person's dislike of his pain an integral part of the pain itself, is this: they see that pains are unpleasant, and they identify their unpleasantness with the sufferer's dislike of them. The resulting view of pain is defective, however, for it entails that whenever a person does *not* dislike what would otherwise be called a pain, what he feels cannot be a *pain*.

If injury or any other noxious input fails to evoke aversive drive, the experience cannot be labelled as pain. (p. 434)

This means – once again – that, for example, lobotomy patients cannot be feeling genuine pains, and this seems clearly false.

To conclude, let me return briefly to the question: "What is pain?" I answer: it is a sensation of a certain kind, *R*, that is unpleasant (or worse) – i.e., that is normally disliked (or the object of an even stronger kind of aversion). I doubt whether there is, now, any way to characterize the kind *R* that would cover absolutely every case of pain. Later on, when brain physiology is further advanced, there may be some way of doing this: pain, we might then be able to say, is the sensation produced by such-and-such a kind of brain state. But for now, in our present state of relative ignorance, we shall have to be content to say that pain is the kind of sensation that is caused, typically (or in standard cases), by damaged body tissues or tissues that are threatened with damage.

Princeton University,
Princeton, New Jersey

NOTES

[1] Sometimes when we have a pain – say, a headache – we become distracted for a time, engrossed in something else, and "forget about the pain" or "pay no attention to the pain," as we might put it. Does the pain continue during that period, as these expressions seem to imply that it does? If so, the pain would be an unconscious one during that time. But since we now have no *other* use for the expression "unconscious pain," I think it is more plausible to hold, as we certainly can, that the pain ceases to exist when we "forget about it" or "pay no attention to it." (See Armstrong, D.M.: 1962, *Bodily Sensations*, Routledge & Kegan Paul, London; Humanities Press, New York, pp. 49-52.)

[2] People sometimes say that one can at least define 'pain' *for oneself* by reference to the felt quality of his own pains; but I think Wittgenstein has effectively destroyed that claim. (See his *Philosophical Investigations, passim.*)

[3] Pains can be anything from mildly unpleasant to unendurable; for the sake of brevity, I shall use the term 'unpleasant' throughout to cover this whole range of badness.

[4] Reik, T.: 1941, *Masochism in Modern Man* (transl. by Margaret H. Beigel and Gertrud M. Kurth), Grove Press, New York, p. 191. D.M. Armstrong, among many others, also subscribes to this view. See his *Bodily Sensations*, p. 91 (n. 1).

[5] There are masochists who at least sometimes do not require physical pain in order to achieve orgasm. (See Reik, T.: *Masochism in Modern Man* p. 42. [n. 4].) Throughout this paper I am dealing only with those more usual masochists who always require physical pain.

[6] They also think that it has a third dimension, but this is irrelevant to my present purpose.

BIBLIOGRAPHY

1. Armstrong, D.M.: 1962, *Bodily Sensations*, Routledge & Kegan Paul, London; Humanities Press, New York.
2. Saint Augustine: 1949, *The Confessions of Saint Augustine*, transl. E.B. Pusey, The Modern Library, New York.
3. Gardiner, P.L.: 1964, 'Pain and Evil', *Proceedings of the Aristotelian Society, Supplementary Volume* 38, p. 118.
4. Melzack, R. and K.L. Casey: 1968, 'Sensory, Motivational, and Central Control Determinants of Pain: A New Conceptual Model', in D. Kenshalo (ed.), *The Skin Senses*, Charles C. Thomas Publisher, Springfield, Ill., pp. 423-39.
5. Noren S. and A. Davis: 1974, 'Pitcher on the Awfulness of Pain', *Philosophical Studies* **25**, 121f.
6. Pitcher, G.: 1970, 'The Awfulness of Pain', *Journal of Philosophy* **67**, 481-92.
7. Pitcher, G.: 1970, 'Pain Perception', *The Philosophical Review* **79**, pp. 368-93.
8. Reik, T.: 1962, *Masochism in Sex and Society*, originally published as *Masochism in Modern Man* (transl. by Margaret H. Beigel and Gertrud M. Kurth), Grove Press, New York.
9. Sauerbruch, F. and Wenke, H.: 1963, *Pain, Its Meaning and Significance*, George Allen and Unwin, London, pp. 30 and 35.
10. Sternbach, R.A.: 1968, *Pain: A Psychophysiological Analysis*, Academic Press, New York and London.
11. Sweet, W.H.: 1959, 'Pain', *Handbook of Physiology*, Section 1: *Neurophysiology*, Vol. 1, American Physiological Society, Washington, D.C., p. 461.
12. Trigg, R.: 1970, *Pain and Emotion*, Clarendon Press, Oxford, p. 148.

DAVID BAKAN

PAIN – THE EXISTENTIAL SYMPTOM

I. INTRODUCTION

The attempt to come to an understanding of pain is characteristically met at the doorway by the materialist objection that subjective states can, at best, be allowed only a secondary status in the realm of scientific reality. Somehow one must make one's way past the materialist's flaming sword if any progress is to be made. Yesterday, we were treated to Professor Jonas' brilliant set of arguments on the absurdity of the materialistic position. I would like to add the additional argument – my favorite argument – that the position of materialism characteristically entails a nonmaterialistic assumption [1]. This assumption is that out in that external nonsubjective world there is something more than material, namely, law. Law is, in the materialist's frame of reference, the 'ghost in the machine.' Law is not a property of matter in the primary sense. The materialist, in talking of law, characteristically slips either into an idealism or a subjectivism. In either case he has deviated from a strict materialism. Materialistic science allows itself the free use of mathematics without acknowledging that thereby it has abandoned its fundamental materialistic posture. Mathematics and mathematical relationships are not material. Indeed, mathematics is, at once, the most objective and autistic enterprise that we can engage in. When the materialist allows mathematically formulated law, besides matter, in the universe, he has essentially cheated on his fidelity to materialism. He may not acknowledge his mistress publicly. Or rather, he may acknowledge her publicly but deny that his relationship is improper. Nonetheless, he dwells both with her and his legal spouse (matter). In spite of his avowed position, he has his subjectivity, his intimacy and gratification in a way that he could not have with his otherwise uninteresting, unexciting, and dead matter.

The medical man, whose primary intent is to heal, turns to psychology. He turns to psychology because he is aware that the subjective in his patients is important in the disease and in the healing process. Un-

S. F. Spicker and H. T. Engelhardt, Jr. (eds.), Philosophical Dimensions of the Neuro-Medical Sciences, 197–207.

fortunately, he too often finds a psychology which is itself psychophobic, and which, in its interest in being scientific, has emulated the most vulgar position in science, materialism. He too often finds that psychology denies the very subjective which the medical man is coming to the psychologist for in the first place.

Pain stands in the center of the medical enterprise. Medicine seeks to get rid of pain. Yet when it looks to scientific psychology, behold the main phenomenon is denied as subjective and hence unscientific.

The vulgarity of materialism in the practice of medicine was brought sharply to my attention in a conversation with a woman whose mother was in the hospital. Her mother was in severe pain. A young medical student told her not to worry about her mother's pain because "all that was really happening was just the firing of some neurons." The fact is that pain will not be dismissed because somebody says it is unscientific.

Even where sensationalism is accepted in psychology, as in varieties of studies in vision and hearing, the stress is characteristically on the stimulus and the motor response. But pain as a sensation is different from other sensations. Pain has no distal stimulus outside of the organism that is very important. No red object out there, no wave length of light, no frequency of vibration that is terribly significant. The pain stimulus is generally some insult to tissue, not the insulting agent. It does not make any difference whether one is shot with an arrow or a bullet. The external stimulus is not significant for the phenomenon, as it is in vision or hearing.

Let me make one other point about my beloved discipline, psychology. For the last fifty years it has been possessed of a vigorous denial – which has to be repeated over and over again because what it denies refuses to be denied – of thought, feeling and willing, the original subject matter of psychology. In recent years thought has managed to work its way back a little. But for the most part, feeling and willing have been relatively ignored. When one speaks of pain, however, one is talking largely about feeling, on the one hand, and willing, on the other.

In this presentation, I will try to give some indication of the critical importance of volition in connection with pain.

A cultural point needs to be mentioned. We live in Western Civilization. Western Civilization is an extremely interesting civilization. One of the most interesting things about it is that its dominant image for close to two thousand years has been of a young man suffering the excruciating pain of crucifixion. I think that this fact is important for understanding that civilization. I will not elaborate on it except to allude to a point that I want to make in a little more detail later on. This image had the effect of adding compassion to the image of God by allowing the deity to experience pain. It is, of course, a fantastic theological notion. It affected the whole subsequent development of Western Civilization. But I will leave that.

II. PAIN IN THE INDIVIDUAL CONTEXT

Virtually everyone who has thought about pain has made the distinction between pain-per-se, on the one hand, and its interpretation, on the other. I prefer to reformulate this by talking about pain-per-se, on the one hand, and, allowing a Heideggerian influence, being-in-question, on the other.

Pain-per-se is the sensational aspect of pain. Being-in-question is the most primitive interpretation that an organism can give of it. Being-in-question is an extremely primitive response. It is the primitive recognition that pain indicates danger to the life of the individual organism.

This distinction is, however, a very sophisticated one. It is a methodological distinction that we can make. However, the organism in pain may or may not make it. Indeed the very making of the distinction by an organism in pain bears considerably on how that organism will react to the pain. Where the sophistication to make the distinction is not present, then the pain-per-se and the being-in-question are integral in the pain experience.

The very making of the distinction between pain-per-se and being-in-question should be regarded from a developmental point of view. Differentiation is associated with maturation both physically and psychologically. Making a distinction is a psychological maturational endproduct. I have learned, for example, that helping a young person make the distinction between pain-per-se and being-in-question itself helps in the management of pain. I have found it effective, for example, when a child is bruised, to carefully examine and treat the bruise, and then say, "It will hurt. But you are O.K.," in this way distinguishing pain-per-se from being-in-question.

It is of value also to distinguish normal and anomalous functions of pain. Normatively, pain results from some damage to tissue. The organism is alerted to the danger associated with the tissue damage. The pain leads the organism to bring to bear his voluntary system towards appropriate safety and therapeutic measures.

However, as with so many other defense mechanisms, the normal pain mechanism may become anomalous and not operate to bring about the appropriate functional result. The anomalies of pain are quite parallel to the variety of mechanisms in which some normal homeostatic mechanism works to exaggerate injury to the organism, as Selye has indicated in numerous examples.

Thus, while pain is generally a beneficent mechanism, signaling damage to tissue, it often also "impairs the sufferer's ability to work and to think clearly, prevents his sleep, abolishes appetite, lowers morale, and even destroys his will to help himself survive" ([5], p. 99). Many of the paradoxical situations of a clinical nature may be clarified by recognizing that, as with other homeostatic mechanisms, the mechanisms associated with pain may work either beneficently or injuriously.

Pain is a phenomenon associated with organisms which are complex, in the sense of consisting of a large number of subsystems organized into a larger system, relatively poor in regenerative capacity, capable of consciousness, and possessed of a large sphere of voluntary activity. The functionality of pain derives from its existence in this context.

In the normal operation in this context, pain heralds being-in-question, and increase in the probability of death. It evokes the attention of consciousness. It informs appropriate steps for the voluntary system of the organism towards the goal of modifying the probability of death downwards. It informs the therapeutic steps by providing information about the location and severity of the tissue injury. And the course of changes in pain informs of the success or failure of therapeutic efforts.

The pain mechanism may confront an essential dilemma. This dilemma is between healing the injured tissue and its associated subsystem or sacrificing the subsystem for the sake of the remainder of the organism. This dilemma is more or less acute, depending on the degree of regenerative capacity in the subsystem, the safety and therapeutic prerogatives that may be available, and the degree of adaptive capacity in the remainder of the organism should the injured subsystem be sacrificed.

The decision load associated with the dilemma is passed on to the psyche of the organism, which informs the voluntary system. The magnitude of the decision load depends on clarity concerning the nature of the injury and the safety and therapeutic prerogatives, on the one hand, and the adaptive capacity of the organism, on the other hand.

If the decision load is great, the special pain associated with the psyche, which we commonly call anxiety, is generated. Anxiety mounts in the absence of resolution. Anxiety is salience of being-in-question. Anxiety may itself be pathogenic. The psyche has the burden both of managing the anxiety and bringing to bear the conscious processes on the voluntary system.

The psyche may sometimes nurture pain-per-se, in the hope, as it were, of increasing the information to help resolve the dilemma. It may nurture pain-per-se to mask the pain of the anxiety. It may also nurture pain-per-se as a way of completely suspending the operation of the voluntary system, which is itself often a major therapeutic prerogative. Nurturing pain-per-se may itself be pathological when it is not associated with the best safety or therapeutic prerogative.

III. PAIN IN THE SOCIAL CONTEXT

But man is a social being. There are two important ways in which man's social nature bears on the problem of adequately conceptualizing the nature of pain. The first is that the safety of others may be entailed in the experience of pain. The second is that the decision load in connection with pain can be, and often is, passed on to other people and to their voluntary systems.

Within the normal mechanism, as I have indicated, the decision load is passed on to the psyche of the organism. Within that psyche there is the part commonly called the ego, the self as it is phenomenologically known by the individual.

That ego characteristically distinguishes between self and not self. That distinction is important for resolving the dilemma between therapy for the injured subsystem or sacrifice of the subsystem. For example, the teeth of an individual are normally included within the ego. The ego experiences the teeth as "part of me." However, should a tooth become diseased, it becomes experienced as "the tooth which is hurting me," that is, something alien and even injurious to the ego; and the ego may well address the dentist with instruction to "take it out." The tooth has thus become an *it* to the ego, in preparation for the sacrifice.

This example is one which is relatively easy as far as the decision load is concerned. True, the regenerative capacity for teeth is nonexistent in an adult. But the therapeutic prerogative is patent, and the adaptive problem for the remainder of the organism is relatively negligible. However, the example does make the point that there is a certain amount of flexibility in the boundary of the ego.

Thus, while the normal mechanism of pain is in the service of the organism as a whole, in the human organism it becomes specialized as in the service of the ego and that which the ego regards as being its own domain.

That boundary of the ego, that point of separation between the self and the not self, tends normally to coincide with the envelope of skin that covers the organism. However, such coincidence of boundary of the ego with the skin envelope is hardly regular and invariable. The boundary of the ego may be considerably beyond the skin envelope, or it may be considerably within the skin envelope.

The boundary of the ego is a matter of learning and culture. The nature of the reaction to injury of the body is highly conditioned by the nature of this ego boundary, even allowing that the normal mechanisms associated with reaction to tissue injury are operative. It remains a survival mechanism. But it is survival of the ego, rather than the survival of the body.

Persons who commit suicide characteristically regard their bodies as external to their egos. They engage in violence against their own bodies without experiencing it as violence against the 'self.' For the suicide, the death of the body is in the service of the self. The death of the body fulfills needs of the ego, even at the expense of the body.

The ego boundaries of persons who engage in great acts of military heroism coincide with the boundary of country or nation. Endangering their own bodies is experienced as relatively minor compared to the danger to the country or nation to which they are loyal. They experience their heroism, even though it is at the expense of their bodies, as serving the interests of their egos.

One of the major functions of nationalistic propaganda is precisely to enlarge the ego boundary of the individual, to make it coincide with the particular nation, its welfare and its survival.

Military training seeks to enlarge the boundary of the ego to include the superior officer and to delegate to him the normal command function of the ego. The suspension of the normal command function of the individual in the presence of pain is the effect of military training when it is successful. Manliness in the military sense consists in being able to continue to keep the volitional processes in order under the command of the enlarged ego, while pain increases. The ideal of bearing pain under torture is essentially an ideal of the individual maintaining his larger ego identity with the larger group and not sacrificing the larger group to the survival of his own body as an entity.

Although the deliberate induction of pain in the body is hardly the most effective method, it has been used as a method of 'socialization.' Socialization often means the enlargement of the ego boundary, especially with respect to the command function over the volitionary processes, beyond the envelope of skin, and its individual interests. Children are spanked, and pain is deliberately induced for infractions, toward this end. Political theorists, such as Jeremy Bentham, whose work has had a profound effect on jurisprudence, have regarded pain as an ultimate base for cohesion in the political body.

The capacity of the manifestations of pain in one person to evoke compassion in another is one of the most important ways in which pain and the larger social situation are related.

Pain elicits cries and whimpers, which are universally understood as indicating that the producer of the behavior is in pain. Such expressions of pain are prelinguistic. The cry is the first 'word' of the child with clear meaning. There is probably no sound that the human being can utter which is so demanding of attention. I suspect that the attention-demanding character of the cry is probably the result of evolutionary natural selection. Groups of human beings for whom the cry was not attention-demanding and evocative of compassion did not survive.

I also suspect that there is an inherited mechanism in which the experience of pain in one individual tends to lower the threshold of responsiveness to the signs of pain in others. If this is true, it suggests an evolutionary explanation for the existence of pain, for the mother, in childbirth.

I have indicated that it is part of the normal mechanism to bring to bear the voluntary processes to increase safety and to promote the application of therapeutic measures. The cry of pain evoking the compassionate responses of others works to transfer the invocation of the volitionary processes for safety and therapy to others.

Such transfer of the volitionary function to others also allows the person in pain to suspend his own volitionary processes. Such suspension is itself one of the major therapeutic measures that is available to mankind. Selye has indicated that the attitude of 'surrender' is a major therapeutic device. It works to reduce stress and to reduce subsystem isolation characteristically associated with stress. But this attitude of surrender, which really means surrender of control over the volitional processes, requires confidence that the safety and therapeutic measures are being adequately managed by others.

The development of highly competent medical and nursing care – the development of a cadre of people whose talent for providing safety and therapy is high – is an important development in the history of the compassionate invocation of the voluntary processes of others. I suspect, however, that the financial drain on one's personal treasure, which is often associated with the delivery of medical services, is perhaps itself one of the major countertherapeutic features of modern medical care. For this makes the very condition of being sick a factor to increase stress rather than to reduce it.

IV. SOME PHENOMENA OF PAIN

I would suggest that many of the anomalies of pain are associated with that feature of the social context and that feature of the mechanism, in which there is a transfer of the command function over the volitional processes to others for bringing to bear of safety and therapeutic measures.

I believe that it is precisely in the problematics associated with this transfer that the clinical problematic of the anomalies of pain arise.

I would propose that a major factor associated with persistent pathological pain is the inability of the individual to allow sufficiently the transfer to others of the management of his condition. His reluctance to thus transfer the command function may indeed be quite justified; and insofar as the social context is such as not to allow it fully, the pathological pain is, as it were, 'justified.' If, indeed, in his judgment – and let us allow that the judgment can be unconscious – the compassion and/or the therapeutic and safety resources do not exist, he 'holds on' to his pain. For thus 'holding on' to the pain, however inefficient it may be, is somehow, at the very least, to keep the normal individual pain mechanism intact. The 'held' pain is at least a reminder to continue the search for appropriate safety and therapeutic measures.

I would suggest, for example, that, in cases of intractable pain, where the physician cannot find any 'cause' for the pain, that the very information that the physician cannot find the cause of the pain may itself be part of the dynamic. The existence of the pain operates as a continuing prod to the volitional processes of the patient to continue, at the very least, in his hunt for someone who is wise enough to help him, someone who yet might provide the adequate safety and therapeutic measures.

Insofar as prefrontal lobotomy may be effective in some cases in the treatment of intractable pain, it may well be that the effect of the prefrontal lobotomy that is directly relevant is interference with the functioning of the voluntary system. Thus, the prefrontal lobotomy 'relieves,' as it were, the individual of the burdens that are associated with the existence of the voluntary system in the normal individual pain mechanism.

The effectiveness of analgesics may well inhere in their capacity to reduce the activation of the normal voluntary system. Anaesthetics, of course, reduce the activation level of both the sensory and voluntary systems.

Beecher observed some time ago that a characteristic remark made by patients who have been treated by barbiturates, morphine, or prefrontal lobotomy is: "My pain is the same but it does not hurt me now" ([3], p. 228). I would suggest that in each case the effect is on the voluntary system, separating the voluntary system out of the pain mechanism. Beecher's classic observation of the extraordinarily low manifestations of pain in wounded soldiers who had been removed from combat [2] is understandable in these terms. Having been brought in from combat, and having been assured that the burden of appropriate safety and therapeutic measures had been assumed by others, the burden of command over their own voluntary systems was very much reduced.

Kolb's observation that pain in a phantom limb could be induced by bringing problematic features of the patient's life and adjustment, such as occupation or sexual attractiveness, to the patient's attention might be understood as the effect of the arousal of the voluntary system in the patient [4].

And finally, that placebos do, in point of fact, sometimes clearly operate to reduce the experience of pain may be understood as the result of the interpretation of the patient that someone is confidently assuming the burden of his safety and therapy. I would suggest that the critical variable associated with the working or nonworking of placebos lies directly in whether the placebo administration does or does not function to suspend the operation of the voluntary system of the patient towards increasing his safety and therapy.

V. SUMMARY

Pain is then truly the existential symptom, in that it is a herald that existence is in question. In the normal individual the load is transferred to the ego and its voluntary system. In the normal functioning of the pain mechanism it sets up a decision load to work towards therapy for the affected part or to sacrifice the part and its associated functions for the safety of the remainder of the organism.

It also sets up a conflict with respect to the invocation of the voluntary system. On the one hand, it invokes the voluntary system to produce effective safety and therapeutic measures. On the other hand, the invocation of the voluntary system is countertherapeutic in that it produces stress that sets up isolation in a subsystem of the organism.

Within the social context the command function of the ego may incorporate others into itself and thus has consequences for the pain experience and the management of the voluntary system. The command function may be completely transferred to others, where others may confidently take over the safety and therapeutic voluntary activities so that the functioning of the voluntary system may be suspended.

The manifestations of pain produce compassion in others, where those others can take over the safety and therapeutic role, allowing the suspension of the voluntary processes in the individual.

The most effective therapy is that where fully compassionate and fully adequate measures for safety and therapy are taken over by others and in which the patient is allowed by his confidence to suspend his own voluntary processes until he is well. The most effective therapy takes place when the dependency needs of the affected person are fully and adequately met.

York University,
Downsview, Ontario

BIBLIOGRAPHY

1. Bakan, D.: 1968, *Disease, Pain and Sacrifice*, Univ. of Chicago Press, Chicago; 1971, Beacon Press, Boston.
2. Beecher, H.K.: 1956, 'Relationship of Significance of Wound to Pain Experience', *Journal of the American Medical Association* **161**, 1609-1613.
3. Beecher, H.K.: 1960, *Disease and the Advancement of Basic Science*, Harvard Univ. Press, Cambridge, Mass.
4. Kolb, L.C.: 1954, *The Painful Phantom: Psychology, Physiology and Treatment*, Charles C. Thomas, Springfield, Ill.
5. White, J.C. and Sweet, W.H.: 1955, *Pain: Its Mechanisms and Neurosurgical Control*, Charles C. Thomas, Springfield, Ill.

BERNARD TURSKY

THE EVALUATION OF PAIN RESPONSES: A NEED FOR IMPROVED MEASURES[1]

The title of this session, 'Altered Affective Responses to Pain,' implies that the relationship between pain and response has been well documented. In their presentations, the participants discuss complex psychological, physiological, and philosophical implications of the human pain experience without questioning the need for more accurate pain response information. Therefore, as chairman of this session, I will address my closing comments to the need for and the production of improved pain response measures.

Let me begin with a little anecdote to illustrate my point. Recently I visited the laboratory of a prominent neurologist involved in highly sophisticated nerve conduction research. He also maintains a private practice and is much sought after as a specialist and consultant. We discussed his research, and he proudly described to me in great detail the complexities of his laboratory equipment, which had the appearance of a miniature space program launch facility. He was very articulate about his control of all the physical and experimental variables in his research setup and took pains to explain how carefully he measured and controlled noise in his recording system. Our conversation finally shifted to my interests, and I described some of my ideas about developing methods for evaluating an individual patient's perception of pain. I asked him how he made such evaluations in his medical practice and what special instruments he used for this purpose. His answer was to reach into his pocket and solemnly produce a large safety pin which he identified, with almost the same amount of pride as he had expressed for his complex laboratory instruments, as his major test instrument in evaluating his patient's pain sensitivity. I really have no quarrel with my friend's use of the safety pin; however, as a scientist, I would like to know how sharp it is and with how much force it is applied.

This little anecdote illustrates a paradoxical situation: The medical and engineering sciences, in recent years, have joined forces and spent a great amount of effort and money in the development of research and diagnostic instruments to measure and record human vital functions. Accurate records

S. F. Spicker and H. T. Engelhardt, Jr. (eds.), Philosophical Dimensions of the Neuro-Medical Sciences, 209–219.

are kept of each patient's height; weight; acuity of vision, hearing, and reflexes; as well as the patient's cardiovascular functions and blood chemistry. These records are kept on a continuing basis and are used to aid the physician in his diagnosis of pathological conditions. Therefore, it can be regarded as strange that until quite recently the verbal and behavioral expressions and descriptions of pain which clinicians rely on as proven indicators of pathology have been treated in an almost cursory fashion, and little has been done to develop scales and instruments to evaluate these important indicators of disease.

The definition of the pain experience has been divided into two major components [1, 10]. One, the sensory or perceptual component, has been described as the actual sensation or hurt that the individual feels. The second is the psychological or reactive component which the pain sufferer portrays by a series of verbal or physical responses that are partly controlled by various psychological, cultural, and situational factors.

Can these complex individual expressions of pain be translated into meaningful information by a clinician who may also inject into this translation his own experiences and attitudes toward pain? Does this complex interaction create an insurmountable problem in evaluating the total pain experience? An important task for the pain researcher is to devise methods for evaluating both the sensory and the reactive components of pain. The intensity of the sensory component of the pain experience, like most physical stimuli, can be evaluated by the use of standard psychophysical methods, for example, the method of limits [16], signal detection [3], and magnitude estimation [14, 12, 5]. The reactive component, however, is much more complex, and special efforts must be made to assess the various psychological, physiological, and behavioral factors that influence this psychic reaction to pain stimulation. Finally, these two measures must be combined and their interaction determined so that a final equation can be formulated that describes the individual's total pain experience.

To accomplish this measurement goal, it seems reasonable to move the study of pain into the psychophysiology laboratory where verbal, behavioral, psychophysical, and physiological pain responses can be recorded, and well-controlled experiments using highly regulated pain stimuli and rigorous experimental procedures can be conducted.

Recently there has been increased interest, as is evidenced by several national and international pain meetings (AAAS, San Francisco, 1974, and

New York City, 1975), in information related to research in the evaluation and control of human pain, and several pain clinics and research centers have been organized to bring together medical, physiological, and psychological expertise to clinically investigate and attempt to control human pain and suffering. Progress has been made in the laboratory to develop techniques for the study of human pain; specifically, laboratory stimulation methods have been developed to experimentally deliver highly controlled pain stimuli, and sophisticated evaluation techniques such as signal detection and magnitude estimation have been utilized to evaluate quantitative and qualitative reactions to these controlled stimuli. The effect of a host of psychological variables on these reactions to laboratory pain has also been studied, and laboratory studies have been conducted to evaluate the effectiveness of pain control procedures ranging from placebos and active analgesics to the use of acupuncture, biofeedback, and hypnosis.

It has been argued [1] that experimental pain could not be compared with clinical pain because a major component of the pain experience, the individual's reaction to the pain, could not be duplicated in the laboratory. However, contrary to Beecher's early pessimism, a number of investigators have recently shown that the reaction component of pain can be evaluated and manipulated in the laboratory. Gelfand [6] demonstrated that experimental pain tolerance could be differentially manipulated by permissive and nonpermissive instructions; Wolff et al. [19] supported these findings and demonstrated that there was no effect at lower levels of intensity; and Clark [3] demonstrated that a placebo effect could be used to alter the pain response to radiant heat stimulation but that thermal sensitivity remained unchanged. Ultimately, Beecher himself [2] acknowledged that the use of ischemic pain in the laboratory had experimental validity and more recently has used electrical stimulation techniques to evaluate the effectiveness of analgesics [9].

One convincing study [13, 18] conducted in our laboratory established the fact that subjects brought the same reactions to pain into the laboratory that they displayed in the clinical situation. This study was based on an anthropological theory [20] which held that Yankee (White Anglo-Saxon Protestant) women have a phlegmatic, matter-of-fact, doctor-helping orientation toward pain; Jewish women express a concern for the implications of pain and distrust palliatives; Italian women express a desire for pain relief; and Irish women inhibit the expression of suffering and concern for the implications of the pain. It was hypothesized that these disparate, subcultural attitudes toward pain should be testable in the laboratory and that the reactions of these groups to laboratory pain should closely resemble their attitudes to clinical pain. The results of this study were consistent with Zbrowski's theory; no consistent differences were found between groups at threshold levels of electrical stimulation, but highly significant differences were found at tolerance levels. Italian women showed significantly lower tolerance levels. This result is quite consistent with the finding of a present-time orientation among Italian woman with respect to clinical pain, that is, a focusing on the immediacy of the pain itself. This concern may be contrasted with that of the Jewish subjects who also display distress when in pain but for the future-oriented reason of concern for the implications of the noxious stimulation. In our laboratory situation, where the stimuli carried no implications of future impairment, no activation of this concern would be expected for the Jewish group, and, therefore, no difference would be expected between them and the Yankees and Irish who tend to be undemonstrative with respect to pain.

The results of these and other laboratory studies justify the claim for the usefulness of laboratory methods in the evaluation of pain responses, but it may be argued that these methods are too complex and time-consuming for use by the average clinician. In my opinion there is a drastic need for a common language and a simple evaluation procedure to define and quantify each individual's pattern of pain responsivity. The methodology for obtaining this information should be available to the physician or dentist to enable him, in a short examination period, to record for each individual patient a personal pain perception profile. These evaluation steps should provide information about the patient's threshold for stimulation, his perception of several nociceptive levels of stimulation ranging in intensity from perception threshold to pain tolerance, and finally, an evaluation of his ability to use language to meaningfully describe his pain experience.

Methodology already exists for accurately assessing the items of the proposed pain profile that deal with judgments of intensity of physical stimulation. Sophisticated psychophysical procedures have been utilized in several laboratories to demonstrate a lawful relationship between stimulus intensity and subjective judgment. For example, Clark and his associates have used signal detection procedures to evaluate the separate contributions of an individual's sensitivity and response criterion to painful stimulation.

Studies conducted in his laboratory have utilized these methods to evaluate the effect of placebos, analgesics, and acupuncture on the individual's thermal sensitivity and on his subjective criterion for pain. In my laboratory, magnitude estimation procedures have been utilized to accurately scale and cross-modally validate the power function that describes subjective judgments of the intensity of electric shock. Sternbach and his associates [11] recently employed what they termed "a subjective pain estimate" with which patients suffering from lower back pain rated, on a scale from 0 to 100, the pain they had experienced during a given 24-hour period. This evaluation was compared to an appropriately transformed measure of tolerance for ischemic pain produced by pressure-cuff stimulation. This comparison seemed to reflect the effect of cognitive pain reduction strategies.

Although these procedures have successfully been used to evaluate the intensity of pain, we are still faced with the problem of interpreting and quantifying the most common mode of communication related to information about the pain experience. I refer, of course, to the individual's use of language to describe the various aspects of this pain experience. The language used to describe pain is varied and useful. For example, the use of the word *burning* as opposed to *aching* as a sensory pain descriptor may convey information to the clinician indicating whether the pain felt by the patient is related to a nerve injury or to a more visceral complaint. Despite the universal use of verbal report to describe pain, very little has been done to categorize or evaluate the intensity of pain descriptive words; and, therefore, one of the greatest problems confronting the pain researcher is the development of instruments to evaluate information provided in the form of verbal self-report.

In our laboratory, the terms *discomfort, pain*, and *tolerance* have been used to describe three levels of nociceptive intensity. These expressions have been employed in several studies as indicators of definable levels of pain reaction, and we have been able to demonstrate a high degree of reliability for these measures for a large number and variety of subjects across time and locus of stimulation [17]. Although greatly pleased with the success of these efforts, it became apparent that words like *discomfort* and *pain* are generic in their assessment of nociception; that is, these words do not clearly distinguish between intensity, reaction, and sensation. It is necessary, therefore, to develop word scales that can be used to assess independently the magnitude of each of these pain descriptive categories.

In 1971, Melzack and Torgerson made an effort to separate the use of pain descriptive terms into three pain categories [8]. One hundred and two terms relating to pain were obtained from the clinical literature and separated into three major categories. These were:

(1) words describing the sensory qualities of the pain experience – pricking, burning, or stinging;

(2) words describing the affective qualities of the pain experience in terms of tension, fear, or autonomic properties – sore, tender, or splitting; and

(3) evaluative words describing the subjective, overall intensity of the pain – annoying, discomforting, or excruciating.

Groups of doctors, patients, and students were asked to assess these terms on a seven-point category scale of intensity. A high level of agreement was demonstrated among the judges on the relative intensity of each term. Melzack and Torgerson's efforts constituted a major step in the right direction; however, it would have been of even greater value if such pain evaluative responses had been psychophysically scaled and *then* relative magnitudes established. Recently, in the Laboratory for Behavioral Research at the State University of New York at Stony Brook, we have been conducting a research program devoted to developing cross-modally validated, psychophysical, verbal scales for evaluating the strength of support or approval for political phenomena [7]. In this effort, we utilize magnitude estimation procedures to evaluate sets of adjectives that can be used to describe the importance of and support for political or social events, institutions, or decisions. Magnitude estimation has been defined as the simplest, most direct, and currently the most widely used method of constructing scales with ratio properties. Subjects are presented with a series of test stimuli, one at a time, and are required to render numerical or physical judgments proportional to the magnitude of the sensations evoked by each stimulus. The use of cross-modal matching techniques makes it possible to determine the reliability and the validity of these scales of perceived intensity, and the development of calibration procedures [4] to correct for regression bias makes it possible to produce valid, bias-free verbal scales of opinion related to almost any area of interest.

It was decided to utilize these procedures to develop three sets of pain descriptors that could be used to describe the intensity, reactive, and sensory aspects of the human pain experience. The intensity scale is defined as a measure of *how much the pain hurts* in units of intensity, the reactive scale is defined as a measure of *how the pain feels* in units of reaction, and the sensory scale is defined as a measure of *what the pain feels like* in units of sensation. The selection of descripters for each scale was based on a series of pretests: The first served to sort a large number of descriptors into the three categories, and subsequent tests used simple verbal, magnitude estimation procedures in large classrooms to evaluate each scale. These results were then used to eliminate descriptors that were too variable. These methods provided a number of words in each category that offered the widest range and least variability. The three sets of pain descriptors were then embedded in simple sentences, each beginning with the words "The pain is," and a full-scale study was conducted to construct a cross-modally validated, bias-free scale for each of the three categories of pain descriptors.

The paradigm for cross-modal validation of any opinion scale is the following: Subjects indicate the intensity of their opinions by adjusting the strength of two or more different physical modalities for which there are established power functions so that the strength of response in each modality matches the estimate of the intensity of the opinion stimulus. When the matches for one physical modality are plotted against corresponding matches for a second modality, a power function should result whose exponent is equal to the ratio of the exponents of the two physical modalities. If this relationship is demonstrated, the scale is considered to be valid.

Fifty-six undergraduate students took part in this study. The stimuli presented to the subjects were three sets of slides shown on a rear-projection screen placed in front of the subject. Each subject was seated in a comfortable lounge chair. Mounted on the arm of the chair were a sone potentiometer and a dynamometer. Located in front of the subject under the rear projection screen were three illuminable boxes labeled *Talk, Squeeze,* and *Noise*. The box that indicated the expected response was illuminated two seconds before the slide came on.

Each subject was presented with two of the sets of pain descriptors. The combination of stimulus sets and the order of presentation within each set were randomized across all subjects.

Judgments of magnitude of intensity, reaction, and sensation were estimated separately using three response measures: magnitude estimation, handgrip, and sound pressure. In each modality, the subject produced a response proportional to the magnitude of each stimulus relative to a standard. Number estimates were given verbally, sound level was adjusted by the subject using the sone potentiometer, and a standard 100 kilogram dynamometer was used to produce squeeze responses. A fourth response was added at the end of the experiment when each subject was asked to draw lines proportional to the same pain descriptor statements.

The results of this study are described in detail in Tursky [15]. Briefly, however, for all three pain category scales (intensity, reaction, and sensation), the exponents that describe the handgrip, sound pressure, and numerical magnitude estimation response power functions fall within the 95 percent confidence limits of the expected functions, and the product-moment correlations between response modes in all instances range from .970 to .990. These results demonstrate that pain descriptors can be psychophysically scaled in a manner similar to that used to scale physical pain stimuli.

The geometric means of line production responses drawn by each subject at the end of the session were computed across subjects for each stimulus and plotted against the geometric means of the magnitude estimation responses for the same stimuli. Theoretically, the predicted relationship between these functions should be unity. The slopes of the regression lines for log magnitude estimation and log line production are 1.11 for the intensity scale, 1.08 for the reaction scale, and 1.2 for the sensory scale. The product-moment correlations for these measures range from .993 to .997. This is an important finding because line production and numerical magnitude estimation are relatively simple psychophysical response measures that can easily be utilized in the field or clinician's office to calibrate individual pain descriptor response. Table I shows the calculated, bias-free scale values of the descriptors in each of the three scales. Regression bias correction is achieved by evaluating each magnitude estimation, handgrip, and sound pressure response, as shown in the formula at the bottom of the figure. The scale values for each of the descriptor categories display some interesting differences. One is that the 40:1 range of the intensity scale is much greater than the approximately 7:1 range of the scale values of the reactive and sensory categories. This may be an indication of the effective separation between intensity and both reactivity and sensation – although this effect may be influenced by the choice of descriptor.

TABLE I

Relative scale values for three dimensions of pain

Intensity		Reaction		Sensation	
*Excruciating	227	Agonizing	153	Piercing	113
Intolerable	167	Intolerable	145	Stabbing	109
Very intense	154	Unbearable	128	Shooting	106
Extremely strong	135	Awful	98	Burning	80
*Severe	132	Miserable	97	Grinding	79
Very strong	129	Distressing	50	Throbbing	75
Intense	123	Unpleasant	43	Cramping	67
*Strong	101	Distracting	36	Aching	58
Uncomfortable	58	Uncomfortable	35	Stinging	50
*Moderate	50	Tolerable	23	Squeezing	46
*Mild	23	Bearable	23	Numbing	40
*Weak	15			Itching	25
Very weak	10			Tingling	17
*Just noticeable	8				
Extremely weak	8				

Scale values were derived from the following formula:

$$\psi(S) = (ME^{1/n_1} HG^{1/n_2} SP^{1/n_3})^{1/3}$$

Another interesting point is the scale relationship between the descriptors 'uncomfortable' and 'intolerable' which were embedded in both the intensity and reactive descriptor scales. Despite the fact that the two scales were judged by different groups of subjects, the scale distance between the two words is almost identical for both categories, although the ratio between them is greater in the reaction scale than in the intensity scale. These results demonstrate that our assumption of the generic pain descriptor qualities of these words is reasonable.

It must be pointed out that the described research is only a first step in the psychophysical evaluation of pain descriptive terms. The population utilized in this study consisted of college students; further research must be conducted using younger, older, and clinical populations to establish whether universal scale values for these words can be established and how they are affected by clinical pain symptoms. At the present time, work with clinical populations of chronic pain (migraine headache and arthritis) sufferers is planned.

The participants in this session have pointed out that pain plays an important role in the human experience. It influences directly and indirectly the conduct of social and psychological behavior. It is therefore mandatory that any serious discussion of pain take into consideration the response measures that have been used to evaluate that pain experience. These comments are offered as one approach to the development of useful pain response measures.

State University of New York,
Stony Brook, New York

NOTE

[1] Research supported by N.I.M.H. Grant No. MH 22296-01, 02, and 03, and N.S.F. Grant No. S041125.

BIBLIOGRAPHY

1. Beecher, H.K.: 1959, *Measurement of Subjective Responses: Quantitative Effects of Drugs,* Oxford Univ. Press, New York.
2. Beecher, H.K.: 1966, 'Pain: One Mystery Solved', *Science* **151**, 840–841.
3. Clark, W.C.: 1969, 'Sensory-Decision Theory Analysis of the Placebo Effect on the Criterion Pain and Thermal Sensitivity', *Journal of Abnormal Psychology,* **74**, 363–371.
4. Cross, D.V.: 1974, 'Some Technical Notes on Psychophysical Scaling', in H. Moskowitz, B. Scharf, and J.C. Stevens (eds.), *Sensation and Measurement: Papers in Honor of S.S. Stevens,* Reidel, Dordrecht-Holland, 23-36.
5. Cross, D., Tursky, B. and Lodge, M.: (in press), 'The Role of Regression and Range Effects in Determination of the Power Function for Electric Shock', *Perception and Psychophysics.*
6. Gelfand, S.: 1964, 'The Relationship of Experimental Pain Tolerance to Pain Threshold', *Canadian Journal of Psychology* **18**, 36-42.
7. Lodge, M., Cross, D., Tursky, B. and Tanenhaus, J.: (in press), 'The Psychophysical Scaling and Validation of a Political Support Scale', *American Journal of Political Science.*

8. Melzack, R. and Torgerson, W.S.: 1971, 'On the Language of Pain', *Anesthesiology* **34**, 50.
9. Smith, G.M., Parry, W.L., Denton, J.E. and Beecher, H.K.: 1970, 'Effect of Morphine on Pain Produced in Man by Electric Shock Delivered Through an Annular-Disc Cellulose Sponge Electrode', *Proceedings of the 78th Annual Convention of the American Psychological Association,* pp. 819–820.
10. Sternbach, R.A.: 1968, *Pain: A Psychophysiological Analysis,* Academic Press, New York.
11. Sternbach, R.A., Murphy, R.W., Timmermans, G., Greenhoot, J.H. and Akeson, W.H.: 1974, 'Measuring Severity of Clinical Pain', in J.J. Bonica (ed.), *Advances in Neurology*, Vol. 4, *Pain*, Raven Press, New York.
12. Sternbach, R.A. and Tursky, B.: 1964, 'On the Psychophysical Power Function in Electric Shock', *Psychonomic Science* **1**, 217-218.
13. Sternbach, R.A. and Tursky, B.: 1965, 'Ethnic Differences Among Housewives in Psychophysical and Skin Potential Responses to Electric Shock', *Psychophysiology* **1**, 241-246.
14. Stevens, S.S., Carton, A.S. and Shickman, G.M.: 1958, 'A Scale of Apparent Intensity of Electric Shock', *Journal of Experimental Psychology* **56**, 328-334.
15. Tursky, B.: 1975, 'The Pain Perception Profile: A Psychophysical Approach', Paper presented at the Annual Meeting of the American Association for the Advancement of Science, January 29-30, 1975, New York, New York.
16. Tursky, B. and Greenblatt, D.J.: 1967, 'Local Vascular and Thermal Changes that Accompany Electric Shock', *Psychophysiology* **3**, 371-380.
17. Tursky, B. and O'Connell, D.: 1972, 'Reliability and Interjudgment Predictability of Subjective Judgments of Electrocutaneous Stimulation', *Psychophysiology* **9**, 290-295.
18. Tursky, B. and Sternbach, R.A.: 1967, 'Further Physiological Correlates of Ethnic Differences in Responses to Shock', *Psychophysiology* **4**, 67-74.
19. Wolff, B.B., Krasnegor, A. and Farr, R.S.: 1965, 'Effect of Suggestion Upon Experimental Pain Response Parameters', *Perception and Motor Skills* **21**, 675-683.
20. Zbrowski, M.: 1952, 'Cultural Components in Responses to Pain', *Journal of Social Issues* 8 16-30.

JEROME A. SHAFFER

PAIN AND SUFFERING[1]

> "One of our mothers had this to say: 'Sure I had pain, but at no time did I suffer.' "[2]

The study of many sorts of psychological phenomena has been hindered by Behaviorism, the identification of psychological states with the behavior which results from these states. This is particularly true in the case of our topic today, pain, where we have an essentially inner, subjective phenomenon. The *identification* of pain with publicly observable pain behavior encourages lumping together importantly different sorts of behavior, for example, avoidance behavior, expressions of pain, verbal descriptions of various sorts, expressions of negative affect, and the like. These are all treated as behavior of one unified sort, 'pain behavior,' and therefore appropriate for the operational definition of pain. This behavioristic approach has led to ignoring important distinctions within the domain of pain phenomena and has prevented advances in our understanding of such phenomena. We are now beginning to investigate, without embarrassment, the more subtle features within the inner phenomena, even where we lack precise behavioral criteria which provide publicly observable evidence for the presence or absence of these inner features and the degree of which they obtain, although we still would not consider the job done until we had developed such criteria or at least come to understand why we could not.

All three of our speakers take a close look at the inner subjective phenomenon of pain. All distinguish between the *sensation* of pain and certain *inner responses* to that sensation. Professor Pitcher distinguishes between the sensation of pain and the unpleasantness of the sensation. Professor Bakan distinguishes between the sensation of pain (which he calls "pain-per-se") and the "interpretation" of the sensation. Professor Tursky distinguishes between the sensation and the reaction to it. Having made their distinctions, they observe that, typically, there is a strong correlation between the sensation of pain and the response that they have distinguished from it. But they argue that the sensation of pain is not *universally* correlated with the response, either in degree or in kind, and

S. F. Spicker and H. T. Engelhardt, Jr. (eds.), Philosophical Dimensions of the Neuro-Medical Sciences, 221–233.

each points to certain cases, e.g., masochism, lobotomy, ethic differences, etc., in which anomalous relationships occur. Their task, then, is to explain these anomalies.

Pitcher's approach is primarily conceptual. It looks as though there is a *contradiction* in the idea that a sensation of pain could be anything other than unpleasant in some degree or other. Yet in the case of masochism, we seem to have pains which are liked, and in the case of lobotomies we seem to have pains which are not disliked. How can one like, or at least not dislike, what is unpleasant, asks Pitcher.

Pitcher argues that the masochist does indeed feel pain and does indeed dislike pain. He dislikes his pain in the sense or respect that he has a "spontaneous, unreasoned desire that it stop." However, the masochist also likes his pain in the sense or respect that he has an even stronger spontaneous desire that it continue (for the sexual excitement it arouses in him). So the liking wins out, as it were, over the disliking and leads him to seek pain, to be sorry if it stops, and to say he enjoys it. So far no contradiction.

But Pitcher also wants to say that pain is unpleasant, and there would seem to be a contradiction in the idea that one could like something unpleasant. Not so, says Pitcher, because 'unpleasant' does not imply 'not liked by *all* people' but only 'not liked by *most* people.' So pain is unpleasant, i.e., not liked by most people, but that is compatible with its being liked by some few people, e.g., masochists.

Pitcher's analysis of masochism is admirably subtle and complex. But I am not sure that, in the end, I understand his solution. To be sure, there is no logical contradiction in a person's having two simultaneous spontaneous desires, one that the pain continue and the other that the pain stop. But I must confess that I have no idea what such a psychological state could be. I can imagine such desires alternating but not occurring simultaneously.

Pitcher does later give a further account (which he seems to think is the same account but seems to me to be different), namely, that the masochist likes his pain for a certain reason and dislikes it for another. As I understand these expressions, they means that he has some reason for liking them, some reason for disliking them. But in such cases, we can come to an on-balance conclusion; all things considered, one reason is stronger than the other, or else they cancel each other out. On balance, we like the thing, dislike it, are neutral or indifferent about it, or we vacillate. In the case of the masochist, the reasons for liking pain presumably win out. On balance, he likes his pain. Therefore, it is misleading to describe the situation as Pitcher does as one in which the masochist *both* likes and dislikes his pain. I conclude, therefore, that Pitcher's account of masochism does not hold up.

I do not think, however, that masochism should be central to our concerns here. Masochism is understood only to a very slight degree as a psychological phenomenon and even far less as a neuro-medical phenomenon. I therefore turn to what I consider to be of particular interest to us, the case of lobotomy. Pitcher thinks that his account of masochism also illuminates lobotomy cases, where we have patients who say that the pain is about the same but they no longer mind it. If we were to apply what Pitcher says about masochism, we would have to say that although these patients have "a spontaneous, unreasoned desire" that the pain stop, it is exactly balanced by a contrary spontaneous desire that it continue, thus leaving them in a neutral, indifferent state. Now it is *possible* that such be the case, but there is no evidence that it is. No, the case with the lobotomized patient would seem to be one in which there is simply no longer the "spontaneous, unreasoned desire" that the pain stop. But how could there be a pain, one wants to say, which did indeed hurt but for which there was not such a spontaneous desire? Nothing in Pitcher's account would prepare us for that. Pitcher does indeed think that his account of masochism prepares us for the lobotomy case. He thinks that by seeing how a person could *both* like and dislike a pain we can see how a person could *neither* like nor dislike a pain. But there is no connection that I can see. Pitcher takes it as natural that a person should dislike pain and explores various theories to explain why, in the masochist's case, there should also be a liking of pain. In the case of the lobotomy patient, we need a theory to explain why he does not dislike his pain. Pitcher provides no account of that.

Bakan, as we have seen, makes a distinction between the sensation of pain ("pain-per-se") and the anxiety which may arise from the "interpretation" of the sensation. This "interpretation" concerns what Bakan calls "being-in-question," namely, the degree of danger or threat to the life of the organism which the sensation portends. Bakan characterizes anxiety as "the special pain associated with the psyche." (This suggests that he is distinguishing two kinds of pain, bodily pain vs. psychic pain, although I am not sure how this further claim fits in with the earlier claim that it is the *sensation* of pain which is pain-per-se.) As I understand him, Bakan is suggesting that the sensation of pain poses the problem to the organism – What is to be done to meet this threat to my life? The degree of anxiety produced is a function of the difficulty in coming up with a satisfactory answer. (All of this may be unconscious.)

In particular, Bakan focuses on the social aspects of pain. The very distinction between the self and the nonself may enable the individual to deal with his pain by drawing the boundary more narrowly or more widely than his own actual body. Furthermore, the individual may relegate decision-making to others, and this surrender of control may result in a reduction of anxiety. Bakan suggests that some of the anomalous cases where pain is not minded or considerably reduced (lobotomies, combat wounds, placebos) may be cases in which control and decision-making is given up, and some cases of intractable pain may be caused by an inability of the patient to give up control.

In the course of his paper, Bakan makes a number of distinctions which seem to me to be of great importance. First, he distinguishes the sensation of pain from the individual's "interpretation." In my own account which follows, I shall call this the difference between the sensory aspect and the degree of suffering. Second, he distinguishes between the sensation of pain and its "manifestations"; I call the latter the *expressions* of pain. Finally, he calls attention to the volitional aspects, the what-is-to-be-done issue; I call this the *avoidance tendencies*. All of these seem to me to be important elements in the full story.

Professor Tursky's paper summarizes his extremely important recent empirical research on the topic of pain. He has pioneered in separating out the various dimensions of the pain experience and in providing scales by which subjects can estimate magnitudes. The remarks I shall be making in the bulk of my paper are, I hope, consistent with and in the spirit of Tursky's work. They are, in fact, what I take to be a tidying up and elaboration of the conceptual scheme with which Tursky is working.

I shall now set out systematically the various features which I believe to be associated with pain (see Figure 1). It seems to me that there are six features which are related to pain, and they fall into three groups as follows: Sensory, Affective, and Behavioral aspects. (I leave out the *temporal* dimension, concerning duration and frequency, since it applies to all six aspects.)

PAIN

A. Sensory Aspect	(1) *Location*:	where the pain is (e.g., in both arms, all over)
	(2) *Quality*:	what it feels like (e.g., piercing, burning)
	(3) *Intensity*:	how much the pain is (e.g., very mild, severe, extremely intense)
B. Affective Aspect	(4) *Suffering*:	how much the pain is minded (e.g., not at all, bearable, agonizing, excruciating, unbearable)
C. Behavioral Aspect	(5) *Expression* of pain:	how the pain is shown (e.g., writhing, moaning, crying, complaining)
	(6) *Avoidance* tendencies:	strength of drive to seek relief (e.g., demanding analgesics, attempting suicide)

Fig. 1.

Tursky does not deal with my first item, *location*, in his paper. What I call the *quality*, he calls the sensation. *Intensity* is the same for the both of us. What I call *suffering*, he calls the reaction. He does not deal directly with the *expression* of pain or the *avoidance* tendencies, but he might do so in other work. In his classification, there is some overlap which I would think undesirable. Thus, in his Table I, 'intolerable' appears as both an intensity magnitude and a reaction magnitude; I would confine it to the reaction magnitude, a degree of what I call *suffering*. Also, I would not put 'excruciating' in the intensity dimension but in the reaction or *suffering* dimension.

There is considerable uniformity among most people so far as the sensory aspect is concerned, but there are various sorts of exceptional cases. With respect to location, there are those who are insensitive to pain in their bodies, as well as those who locate pain outside their bodies, e.g., phantom limb pain ([5], ch. 8). As for intensity there are those whose general threshold for pain is higher than normal, e.g., some mental retardates ([5], pp. 96-97). And there are those who are congenitally totally insensitive to pain ([5], ch. 7). Analgesics, placebos, and hypnosis also differentially affect the sensory aspect in varying degrees, for various individuals, under various circumstances.

There are also various sorts of exceptions to the norm under the affective and motivational aspects. It is a familiar fact that expression of pain varies from culture to culture (Lawrence of Arabia reported that he was able to maintain his disguise as an Arab by crying out in typical Arab ways when he was tortured by the Turks.) It is also a familiar fact that expressions of pain vary as well from individual to individual within a culture. It is more surprising to discover that there are exceptions to the standard correlations between degree of pain as a sensory fact and the degree of suffering involved or degree of desire to reduce the sensory element. When one first hears of cases of people who say the pain is just as great but they now mind it much less (e.g., lobotomy patients) or say that it hurts but they seek it out and try to prolong it (e.g., masochists), one is inclined to doubt one's ears or else insist that it cannot be literally true. One is inclined to think that where there is pain there will be negative affective and motivational responses and the greater the intensity of pain, the greater those negative affective and motivational responses. By and large these correlations do usually obtain, and the fact that they do usually obtain has made it difficult and controversial to make the very distinctions here made. Even where the distinctions have been made, it has nevertheless often been alleged that there are necessary connections between them; so that, it is alleged, it cannot be pain unless it is minded or expressed or there are some efforts to avoid it.[3]

Given the dimensions of pain phenomena set out above (see Figure 1), we can represent the anomalies as a series of exceptions to commonsense beliefs about pain. The following schema is a summary table of the rest of the paper:

Commonsense Beliefs	*Refuted By*
Intensity is proportional to suffering	Lobotomies, natural childbirth, war wounds, LSD
Intensity is proportional to expression	Cultural differences
Intensity is proportional to avoidance	War wounds, natural childbirth, masochism, LSD
Suffering is proportional to expression	Cultural differences
Suffering is proportional to avoidance	Masochism
Expression is proportional to avoidance	Lobotomies

Consider what I refer to here as suffering. It is, I think, what Professor Pitcher has in mind by unpleasantness (in an earlier paper he called it the *awfulness* of pain), what Professor Bakan has in mind by the stress or anxiety which arises from the perceived danger ('being-in-question'), and what Professor Tursky calls the 'reaction' to the sensation. I think we all mean here a range of responses which at the upper end consists in a degree so great that it could not be augmented, where we would say "it is more than the individual can bear," and which at the lowest level consists of indifference, of not minding it at all, of utter neutrality, with levels in between in which the pain is bearable, unpleasant, pretty bad, awful, etc. This range of evaluative terms is not restricted to pain, of course; the concept of suffering is a wider concept than that of pain. Not only is there mental suffering, produced by fear, bereavement, failure, insults, etc., but other sensations besides pain produce suffering. The most fearful suffering I myself have ever experienced was a case of seasickness – but I would not say I had been in *pain*. Extreme itching might also produce much suffering even though it does not involve pain. So not all cases of suffering are cases of pain. But are all cases of pain cases of suffering? Probably not a very short-lived pain or a slight but habitual pain. But, more importantly, the pain in natural childbirth, where it is an expected element in a much desired process and a welcome sign of a much desired outcome, is probably not a case of suffering.[4] I am not sure what to say about the case of the masochist here. I agree with Pitcher that the masochist may well experience genuine pain, but does he suffer? Perhaps it varies from masochist to masochist. It is not inconceivable that, as in the natural childbirth case, we have here a necessary element in a much desired process, with a much desired outcome which depends on the pain, hence pain without suffering. Alternatively, it is a case in which the suffering is there, but with no tendency to avoid it and actually a strong tendency to seek it out, thus a counterexample to the alleged universal correlation of pain and suffering with avoidance efforts.

The most dramatic case of the breakdown of the correlation of pain and suffering is that of the lobotomy patients. As Bakan points out, Beecher quotes, "My pain is the same, but it doesn't hurt me now" ([1], p. 228). Others report that a patient "continued to perceive pain but was no longer troubled by it and mentioned it only when questioned" ([6], p. 69). Two professors of surgery, Roberts and Vilinskas, say that a lobotomy "produces indifference to chronic pain."[5] Melzack says that "typically these patients report after the operation that they still have pain, but it does not bother them" and goes on to point out that "it is certain the sensory component of pain is still present because these patients complain vociferously about pinpricks and mild burns."[6]

There is some disagreement about this interpretation. Merskey and Spear ([4], p. 23) are of the opinion that lobotomy patients do not feel pain, but they seem to hold this opinion as a result of their general account of pain as *by definition* unpleasant. They offer no specific evidence that lobotomy patients feel no pain and do admit, for example, in the case of the removal of stitches, that "the reaction of the lobotomized patient to noxious stimuli may be vigorous, or increased, although his anxiety or concern over it appears less" ([4], p. 159). I would, myself, take this as disconfirming their claim that there is a necessary connection, as a matter of the definition of pain, between having pain and finding it unpleasant.

LSD seems to have a similar effect. After taking LSD, a terminal cancer patient is reported as saying:

> And what about the pain. I suppose I'll be a baby about it again. Right now, the pain is changed. I know that when I pressed here yesterday, I had an unendurable pain. I couldn't even stand the weight of a blanket. Now I press hard – it hurts – it hurts all right – but it doesn't register as terrifying. It used to throw me and make me beg for another shot. [2]

This would seem to me to mean that what has changed is not the intensity of the pain so much as the degree of suffering the pain produces.

It is of great importance for more research to be done on the relation between pain and suffering. Why does pain usually produce suffering? What neurological mechanisms are involved? What psychological mechanisms are involved? How can the unpleasantness of pain be reduced?

Consider the beautiful experiment done by Tursky and Higgins.[7] When nearby tactile stimulation was added to electrical shock, there was no change in the reports by the subjects of the magnitude of *intensity* for comparable electrical inputs, but it took greater electrical inputs to elicit the same negative affective responses in terms of the magnitude of discomfort and the ability to tolerate the shock. Here we have a clear experimental distinction between the intensity of the pain and the degree of suffering; tactile stimulation was able to provide for a diminution of suffering under similar intensity of pain. It hurt as much, but they minded it less, to use the words of lobotomy patients. A similar result appears to occur when a person has control over the occurrence of the shocks. "Not having control or relinquishing control results in a reduction in the subject's tolerance for aversive stimulation."[8] This result, by the way, seems to count against Bakan's hypothesis that surrendering control diminishes suffering.

When we come to the relation of pain to the *expressions* of pain, a host of questions arise. What are the expressions of pain? How and why do they vary from culture to culture? What are their functions? (Professor Bakan points to their social function in eliciting the attention and help of others.) Do the expressions of pain in turn affect the pain that produces them? Under what circumstances are they a matter of voluntary control or involuntary response? How specific are they to *pain*? (Consider how the expressions of sexual pleasure, particularly orgasm, resemble the expressions of pain. I am told that in pornographic movies, one reason orgasms are simulated is that if actual orgasms were filmed, the audience would take the subject to be in considerable pain.) What is the effect on the patient of the inhibited or uninhibited expression of pain? What is the relation of expression of pain to intensity, to degree of suffering, to avoidance efforts? We know from Tursky's work and that of others that cultural and ethnic factors play a role in determining the degree to which a person in pain will express that pain. Lobotomies seem to have the effect of increasing demonstrativeness in the expression of pain while decreasing the amount of suffering (as indicated by subjective report) and while decreasing the tendencies to avoid or reduce pain intensity (as indicated by requests for analgesics) ([3], pp. 316-322). In short, there is no simple correlation between expression of pain and the other dimensions of pain.

Similar questions arise for avoidance efforts (as well as the cases in which pain is actively sought out or under voluntary control). Melzack mentions the way in which toothaches sometimes tend to disappear as we enter the dentist's office. (I find I can often get rid of a headache simply by *deciding* to take an aspirin.) So there is the general issue, how avoidance efforts affect the other elements. And there is the set of questions, how avoidance tendencies vary with changes in the other elements. It would seem likely that high intensities of pain do not always correlate with avoidance tendencies. But is there a closer connection between suffering and avoidance tendencies? When the pain is unbearable, will the subject do anything to avoid it? Is that what 'unbearable' means? [9] Or rather, as I would be inclined to think, isn't it that 'unbearable' means something like 'will faint or crack up if it continues' so that there can be those who even under unbearable pain will not recant or tell us where the plans are. Our data are very skimpy here, for obvious reasons.

If indeed we are justified in distinguishing between the sensation of pain and the affective and motivational features connected with it, what should be *our own* affective attitudes to the various anomalous cases? Should we feel pity? Should we try to ameliorate the situation, for example, by administering an analgesic? In the case of masochism, where, let us say, there is pain, suffering, and the expression of pain in moaning or the like, but where there is intentional pursuit of painful stimuli, should we interfere, administer an analgesic, for example? I should think not. That is the sort of meddlesome do-gooding up with which no self-respecting masochist should put. Similarly, in the cases of natural childbirth, combat casualties, and lobotomy patients, there would seem to be no further need for analgesics than are requested by the patient. In short, it would seem that our attitude of pity and the belief we should intervene is a result of our belief there is suffering rather than a result of our belief there is pain.

Lobotomies seem to leave the pain and only remove the suffering. How about thalamotomies which remove both the pain and the suffering? Disregarding the side effects (which in the case of lobotomies are important negative considerations), can we say that a procedure which eliminates *both* pain and suffering is superior to one which eliminates only the suffering?

Sternbach believes so, saying, "In many respects, thalamotomies are preferable to prefrontal leucotomies for the relief of chronic severe pain, since it is now well established that the latter procedure frequently only eliminates the complaints of pain, but not the feeling of pain" ([5], p. 35). Of course, if lobotomies eliminate only *complaints*, taken as expressive behavior, but allow the suffering to continue, they would be no better than curare which, by producing paralysis in the patient, would also eliminate complaints. Lobotomies eliminate the need to complain by rendering the patient indifferent to the pain (which Sternbach himself admits).[10] If the patient is indifferent to pain, no longer minds it, then I do not see why one should prefer a treatment which, in addition, takes the pain away. That seems superfluous. If a lobotomy patient developed lesions in the thalamus which removed the lingering pain and then expressed relief, then it would show he had minded the pain after all, contrary to our hypothesis.

Ordinarily, then, the sensation of pain induces suffering, which induces expression of pain and avoidance tendencies. In such cases we are under the obligation to do what we can to reduce or eliminate the pain. Where there is no suffering (lobotomies, natural childbirth, war wounds) or where there is suffering but no interest in avoiding it (masochism), we are under no such obligation to reduce or eliminate the pain.

Finally, in the light of the distinctions we have drawn concerning pain, what sense are we to make of the masochist? There are a number of possibilities which may or may not correspond to actual varieties of masochists. There might be those who, in their sexual excitement, do not even feel the pain at all; as Pitcher points out, it might be the scene and the fantasies it evokes which really turn them on. Then there might be those who feel the painful sensations but do not suffer, that is, do not mind the pain. But what are we to make of the alleged claim that they (some at least) positively like or enjoy the pain? One thing this could mean is that the masochist does indeed mind his pain, does suffer, but nevertheless seeks it out rather than avoids it, for reasons which are, at present, not well understood. But there is another thing it could mean. We have so far only discussed the negative affective dimension, suffering, which goes from the neutral, zero point, indifference, not minding something, to the maximum point, finding it unbearable. But there does exist a positive affective dimension as well, which goes from indifference through various degrees of liking it, to various degrees of loving it. Let us call this positive affective dimension *enjoyment*. It would be unusual, but not impossible, that under certain circumstances someone should have a severe (in intensity), piercing (in quality) pain and enjoy it (in a positive affective response). What would be impossible, on my account, would be that someone should *enjoy suffering*, enjoy a pain he at the same time minded, for that would involve simultaneously a positive and negative affective response, a contradictory state of affairs, the way we have set it up. As for the groans, winces, and cries which Pitcher, admitting that it is only by hearsay, ascribes to the masochist, it is not obvious, as Pitcher seems to assume, that they are expressions of pain, since, as we have mentioned above, expressions of sexual gratification are often similar to expressions of pain. But even if they are evidence of pain, it does not follow that they are evidence of suffering, for, as we have seen, the connection between the expression of pain and actual suffering is quite complex. So even when there are groans, winces, and cries, they may stand for very different sorts of things in different cases. In short, our various distinctions concerning the phenomena of pain open up a variety of possibilities. It is only when we begin to get more empirical data that we will be able to tell which of these possibilities actually obtain. And given the bewildering complexity of the data, not just in the case of masochism but in the entire spectrum of pain phenomena, I think it is safe to say that any serious investigator has a great deal of suffering ahead of him. Perhaps we can find some able masochist to take on the job.

University of Connecticut,
Storrs, Connecticut

NOTES

[1] Preliminary work on this paper was done while on a Senior Fellowship from the National Endowment for the Humanities.

[2] Chabon, I.: 1969, *Awake and Aware*, Delacorte Press, New York, p. 13. A copy of this book was presented to me at the Symposium by Walter Vesper of Meharry Medical College. I am most grateful to him for calling my attention to the natural childbirth and psychoprophylaxis literature.

[3] Melzack, R.: 1973, *The Puzzle of Pain*, Basic Books, New York, says: "If injury or any other noxious input fails to evoke negative affect and aversive drive (as in the cases described earlier of the football player, the soldier at the battle front, or Pavlov's dogs) the experience cannot be called pain." (pp. 46-47)

[4] Natural childbirth, or more accurately, psychoprophylaxis, also seems to reduce the *intensity* of pain sensations through exercises, relaxation techniques, etc. But it would also seem to reduce the *suffering*, especially at the moment of birth where the intensity of the sensation of pain is presumably fairly uniform. See Chabon (n. 2).
[5] Roberts, M. and Vilinskas J.: April 1973, 'Control of Pain Associated with Malignant Disease by Freezing: Cryoleucotomy', *Connecticut Medicine* 37, 185. Professor Roberts was a participant in the Symposium.
[6] Melzack (n. 3), p. 95. This would appear to contradict his earlier remarks cited in note 3.
[7] Reported in Tursky, B.: 1974, 'Physical, Physiological and Psychological Factors That Affect Pain Reaction to Electric Shock', *Psychophysiology* 2, 107-108.
[8] *Ibid.*, p. 106.
[9] E.G. Boring, for example, while accepting the distinction between pain as a sensation and the unpleasantness of pain, identifies the unpleasantness with "something to be avoided, rejected, escaped from, or terminated" (Hardy, J.D., Wolff, H.G., and Goodell, H.: 1967, *Pain Sensations and Reactions*, Hafner Press, New York, p. v.).
[10] Sternbach, R.A.: 1968, *Pain*, Academic Press, New York, pp. 97-98. In this passage, Sternbach identifies *indifference* to pain with "fail[ure] to withdraw from or avoid pain stimuli" (p. 97). I distinguish the two. Sternbach himself admits that lobotomy patients, indifferent to pain, "usually manage to avoid injury as insensitive patients do not" (p. 98), thus contradicting himself and supporting our distinction between the affective dimension and the avoidance-tendency dimension.

BIBLIOGRAPHY

1. Beecher, H.K.: 1960, *Disease and the Advancement of Basic Science*, Harvard Univ. Press, Cambridge, Mass.
2. Cohen, S.: September 1965, 'LSD and the Anguish of Dying', *Harpers Magazine* 231, 77.
3. Hardy, J.D., Wolff, H.G., and Goodell, H.: 1967, *Pain Sensations and Reactions*, Hafner Publishing Co., New York.
4. Merskey, H. and Spear, F.G.: 1967, *Pain*, Bailliere, Tindall and Cassell, London.
5. Sternbach, R.A.: 1968, *Pain*, Academic Press, New York.
6. Wolff, H.G. and S. Wolf: 1948, *Pain*, Charles C. Thomas, Springfield, Ill.

SECTION VI

Round Table Discussion

THE FUNCTION OF PHILOSOPHICAL CONCEPTS IN THE NEURO-MEDICAL SCIENCES

ROUND-TABLE DISCUSSION

WILLIAM F. BYNUM

Being neither a philosopher nor a neuroscientist, I can confidently claim that a round table discussion on the function of philosophical concepts in the neuromedical sciences would appropriately find me in the audience rather than at the table. At best my comments will tend to be historical and descriptive, not philosophical and prescriptive. Further, these comments will have a negative aspect, for my own historical researches have convinced me that, historically, the relationship between philosophy of mind and the neurological sciences has not been so fruitful as it might have been. One of the ongoing problems of the neurological sciences has concerned the nature of the nexus (if any) between mental function and neurological function. The history of neurology is littered with statements purporting to prove that this relationship corresponds with certain authors' metaphysical presuppositions. But the tradition of introspective philosophy and philosophical psychology has largely ignored the relational issue, and historically, philosophers by and large have not had to concern themselves with the physical basis of mind. Neuroscientists, on the other hand, have frequently accepted a philosophical account of mental phenomena which left them without an adequate conceptual framework for translating their neurological investigations into practical descriptions of behavior, mental activity, etc. Of course, part of the problem has stemmed from the technical difficulties inherent in brain research. But in other ways these difficulties are a residue of the Cartesian legacy which still casts shadows even on meetings like the present one.

I should like to suggest three problems which strike me as symptomatic of the historical barrier which has existed between philosophers and neuroscientists. Their importance has perhaps diminished but not disappeared in the present century. These problems are as follows: (1) the historical confusion – and sometimes conflation – of questions of vitalism with questions of dualism; (2) the problematic status of the animal mind; (3) the failure to relate philosophy of mind to insanity and disease. It is churlish

S. F. Spicker and H. T. Engelhardt, Jr. (eds.), Philosophical Dimensions of the Neuro-Medical Sciences, 237–268.

perhaps to blame Descartes for subsequent difficulties involving these issues, though it is significant that none of them was satisfactorily resolved within the Cartesian framework. And one major feature of much neuroscience – certainly through the 17th, 18th, and into the 19th centuries – shares with Descartes his fallacy of misplaced confidence: the belief that any characteristic which men possess in common with animals is much more easily understood than any characteristic which men seem to possess uniquely. Within the British context, at least, the precise way in which Descartes expressed this belief – in terms of the doctrine of the animal machine – met with little favor. On the other hand, the moral and theological consequences of Cartesian dualism were espoused by numerous physicians and biologists, who have considered the mind as a thing apart. However, it has proved exceedingly difficult to adhere to a strictly dualistic account of the human mind without either subscribing to the notion of the animal machine or blurring the clarity of Descartes' juxtaposition of mind and matter. Although a love of and respect for animals has permeated the British mentality for centuries, only a minority of British thinkers have felt comfortable with a thoroughgoing doctrine of psychological continuity between men and animals. Even while rejecting the rather awkward Cartesian doctrine of the "ghost in the machine," they were unwilling to abandon the theoretical (and largely theological) underpinning of traditional introspective dualism. At the same time, many physicians and life-scientists have failed to notice that a doctrine of vitalism vitiated (within the context of 18th and early 19th century comparative neuroanatomy) their own beliefs in the ontological distinctness of the human mind. Descartes believed that animal life and behavior was explicable in mechanistic terms. By countering him at this point, British vitalists such as John Hunter, John Abernethy, James Prichard, and others, created a framework out of which a doctrine of psychological continuity could easily emerge.

This interplay between vitalism and dualism deserves more historical attention, since the resultant confusion was integrated into a number of debates during the past three centuries. Cabanis attempted to throw questions of mind into general biology by his famous analogy of thought with bile, an analogy which has satisfied neither philosophers nor scientists, since Cabanis attempted to compare unlike entities. More specifically, William Lawrence in the 1810's borrowed from French biology the notion of an organisational vitalism which would make mental and vital phenomena equally dependent on organic structure. He was accused of materialism, whereas he had merely opted for biological continuity in a special sense (that structure and function are related in the same way in men and animals). Lawrence's work contained a statement of psychophysical parallelism, but, historically, this kind of parallelism has not so much been some halfway house on the road to materialism and reductionism as the abode of those who believed that law was discoverable in all the phenomena of our experience. David Hartley is a good case in point. However, physicians and other life-scientists have had a rich body of information on which to draw for their statements about the relationship between brain physiology and mental function. Their writings have been diluted by appeal to this information, which after all gave some point to the study of the neurosciences. They have not felt able to take advantage of the luxury employed by someone like Locke of discounting at the outset any anatomical or physiological issue. And in their appeal to general biology (and to natural theology) they have often arrived at conclusions different from those of the general philosophical traditions in which they saw themselves. The debate surrounding phrenology provides a rich case study of this issue. I have examined aspects of the debate in my paper to this conference, where I suggested that both phrenologists and their critics relied on an underlying notion of psychological continuity. In the case of someone such as William Hamilton, this reliance was presumably unconscious, for Hamilton's stated intention was to vindicate traditional introspective analysis of mind. The extension of his experimental work to man strongly suggested a more clearly defined notion of animal minds than his own philosophical system could accommodate. His own dualistic and voluntaristic philosophy was an inappropriate setting from which to launch a crude kind of experimental cerebral physiology. Historically, vitalism and dualism have frequently been invoked to serve similar philosophical functions. However, the routine substantiation of these two separate positions involved different assumptions and bodies of evidence, and the simultaneous use of a principle of vitalism and a doctrine of dualism often led to awkward inconsistencies.

The point is that the Cartesian tradition did not provide a model with which British philosophers or scientists were comfortable. Nevertheless, many retained the parts of Descartes' work which suited them. They resisted the notion of some precise form of brain-dependence of human minds, yet they recognised that their science was inadequate to explain the activities of animals in mechanical terms. The compromises they reached were not always clear, particularly as long as the Cartesian mind was equated to the theological soul.

Many of these same problems can also be found in the 18th and 19th century literature on insanity and criminal responsibility. The secular variety of the notion of criminal insanity was difficult to square with the voluntaristic strands of introspective philosophy, on the one hand, and the existing legal codes on the other. Saddled with a belief in the human mind which was simultaneously immaterial and substantial, medical men in the early decades of the 19th century attempted to view insanity in terms of the body. It is, after all, unclear how one would elaborate a secular notion of primary mental disease if the human mind is equivalent to the theological soul. In practice, medical men tended to project the mind as an entity in a way which itself undercut the inheritance of Cartesian categories, by making the operation of the mind (as opposed to the mind in itself) dependent on neurological structures. There is need for much work on the ideological and philosophical underpinning of 19th-century notions of insanity as a prelude to understanding the contemporary polemics of Szasz regarding psychiatry in the West, and the warnings of the Medvedev brothers and others about the situation in Russia. In our post-Freudian obsession with "mental diseases" we may have forgotten the important (though in many cases naive) proposition of the 19th-century psychiatrists who insisted that real disease must be physical. Our division of medical specialties into psychiatry and neurology is Cartesian at its core, although if psychosurgery is the kind of hybrid we may expect from the marriage of physical and mental sciences, perhaps Cartesianism isn't so bad after all. Be that as it may, it is ironic that Szasz's critique of contemporary psychiatry still implicitly retains Descartes' categories of mind, body, and free will, at a time when philosophy has largely escaped the polarities of the Cartesian universe. Psychiatrists might do well to take seriously the alternatives to materialism and dualism posed in the work of philosophers such as Strawson, Stout, Grene, and others. Antony Flew's recent examination of the issues raised by Szasz might also be taken as an encouraging sign that philosophers are becoming more engaged in the practical problems of modern medicine. Psychiatry itself has been impoverished by the historical neglect which philosophers have shown for the abnormal, though of course since Freud, philosophers have not been diffident about entering the psychoanalytical fray. The notion of disease itself, however – social, mental, physical, and perhaps even "embodied" – is still a fuzzy and misused concept which could be refined by philosophical analysis. Whether we shall ever reach a point in which there are only diseases of "persons," rather than diseases of bodies and diseases of mind, is an issue on which I am not competent to speak. If we do, I suspect that it will come through a philosophical reorientation rather than through any new laboratory discoveries.

On the other hand, I am enough of a materialist to be suspicious of an account of minds or persons which does not theoretically allow a complete physical basis. We may agree with Shaffer that brain states and mental states are not identical. But as Shaffer himself suggests, a physical transcription of any mental event – though only a transcription of convenience – is theoretically possible. And I suspect that mental processes are such that an exact molecular copy of myself would have my same desires, intelligence, memories, etc., i.e., it would have the *record* of my experiences even if it had not had those experiences themselves. In that sense, I am my physical self, though in practice it is not particularly useful to think in such terms. One would undeniably lose a great deal in the process of reduction. It is thus debatable whether complete knowledge of the physical basis of the self would provide a humanly satisfying account of a self, but that, I think, is a separate issue from the theoretical possibility of such an account.

SAMUEL H. GREENBLATT

Since I am the only practicing neurosurgeon or neurologist on this panel, it occurred to me that I might have been expected to say something clinical – a reasonable expectation at first blush, but seemingly inconsistent with the practical facts of life. Very few clinicians or experimentalists in our day have had any useful exposure to formal philosophical analysis. Most of them just seem to think their thoughts and do their work with only the vaguest awareness of their state of deprivation. And yet, the work goes on, busier and busier. I suspect that if one were to ask an average neuroscientist about the role of philosophical concepts in his work, the response would be biphasic. The first phase would be non-verbal – a sort of puzzled look slowly evolving into a wry smile. The second phase would be verbal, perhaps markedly so, with the essence of the verbiage consisting in lip service to some vague humanistic ideal.

To the casual observer, it would appear that philosophical concepts have no significant role in the practice to the neuromedical sciences *at the conscious, workaday level.* On the other hand, most of us have assembled here because we share an intense belief that, at some level, this can not be the case. Surely we all use ethical and logical concepts, even if they were not imparted to us through formal study. But ethics and logic are broadly applicable to all of medicine, whereas this conference is concerned primarily with the exploration of those philosophical concepts which are uniquely important to the neurosciences.

In order to seek out this special relationship, let us return to our philosophically naive neuroscientist, in the laboratory or the clinic. We want to ask once more whether or how he might use philosophical concepts specifically related to his own field. The answer will have to be obtained by our own observations of his behavior, because he has already shrugged us off and gone back to work. For the sake of convenience, I will refer to our subject as Dr. Neuro, since that is what the nurses call him, and the name has the advantage of applying to either gender.

It is axiomatic that the central focus of Dr. Neuro's activities is the brain. From one point of view or another, he wants to know how it works, or how it fails to work in some instances. It is equally axiomatic that when Dr. Neuro starts to worry about how the brain works, he does not begin with a *tabula rasa.* He begins instead with a very complex intellectual apparatus. The more visible superstructure of this apparatus is based on repeatedly confirmed clinical and experimental observations. But very often the established facts and theories are insufficient to explain his own observations. As a practical matter, he can not always ignore these gaps. Sometimes he must fill them in as he proceeds or he will soon be bogged down in hopeless disarray. Very well, then, how does he fill in the gaps? Why, of course! He makes assumptions. These gap-bridging assumptions are based on what he perceives to be the most likely pattern of brain events, based on his established knowledge of how the brain works. It is precisely at this juncture where I believe that philosophical concepts begin to intrude upon the active thought processes of our erstwhile neuroscientist.

Whenever any experimentalist or clinician works out a problem, he must bridge gaps. In doing so he naturally depends on some basic metaphysical assumptions, because the process of bridging the gap involves a feeling for the reasonable limitations on the behavior of the system under study. Assumptions about the nature of the real world are called into play in deciding where these limits may reasonably be found. In the case of the brain specialist, these philosophical assumptions take on special importance, because there has been a longstanding philosophical concern with the analysis of mind and its relation to brain activity. I think it is fair to say that among neuroscientists and neuroclinicians the current general consensus holds that there is some kind of intimate relationship between the operations of the mind and the activities of the brain. True enough, if we are pushed against the wall, most of us will readily admit that we are stumped by the problem of Cartesian dualism. But when given a little elbow room, we ignore the problem. We function as integrationists or parallelists or something like that. If this point is admitted, then I submit that in actual practice this is the means by which philosophical concepts intrude upon neurophysiology, because the historical tendency has been implicitly to translate concepts of mind function into concepts of brain function. In Kuhn's terms [3], we might say that Dr. Neuro's paradigm – his working model – is partly constructed with assumptions which derive from philosophical conclusions about the operations of the mind.

The point that I am making, then, is essentially an historical one. It applies not only in the process I have called gap-bridging, but also in our utilization of facts and theories which we take to be observationally well established, because we still have to decide what we will choose to observe. In order to explore this further, let us take a closer look at Kuhn's very useful idea of the paradigm. I defined it briefly above as a "working model," but it is really much more. When a scientific paradigm is widely accepted, it constitutes a very pervasive conceptual framework for those who use it. In other words, it sets conceptual limits on its users. If the paradigm is sufficiently powerful and pervasive, these limitations are not questioned. Indeed, they are not even fully appreciated; they are just assumed.[1] In the case of 20th century neuroscience, I would claim that our basic paradigm is philosophically loaded in a particular way. Hence, our assumptions ought to be at least partially definable by historical analysis.

By now it is, no doubt, obvious that I have in mind some particular historical culprits, and indeed I do. In the English-speaking world, the all-pervasive neurophysiological paradigm is the contribution of Sir Charles Sherrington, whose brilliant work has guided the course of 20th-century neurophysiology since its first decade. Sherrington was blessed with a very extended and fruitful old age. Long before his formal retirement at the age of 78, he undertook to explore the philosophical implications of his own work. The resulting *corpus* is of very great interest to contemporary philosophers and physiologists alike. It has been examined and analysed repeatedly in recent years. But its very existence has inadvertently caused another aspect of Sherrington's work to be neglected – an historical gap, if you will. The philosophical positions which Sherrington outlined in his old age are not necessarily the same as those which he assumed in his youth. In a previous essay I have explored very briefly the intellectual milieu in which Sherrington was raised and educated ([2], pp. 58-61). During the latter decades of the 19th century, the prevailing philosophy of mind in Great Britain was based on the associationist psychology. Clear traces of this influence can be found in the closing chapters of Sherrington's lectures on *The Integrative Action of the Nervous System*,[2] which was published in 1906, but they seem to have disappeared from his later philosophical work [4].

A central theme in associationist psychology and Sherringtonian physiology is the sensory-motor reflex paradigm. When applied in a strict *a*-to-*b*-to-*c* kind of connectionism, it still serves us very well. If *a* and *b* and *c* are conceived very broadly, then it also serves quite well in the Pavlovian sense. Integration of reflexes is another theme which can be traced quite easily from associationism into Sherringtonian physiology.[3] Indeed, the most demanding part of Sherrington's legacy to us is our continuing struggle to understand the nature of integration in physiological terms. It is easy enough to understand how a spinal or even brainstem reflex can be mediated through an arc of 3 cells. It is also not difficult to grasp the idea that the behavior of the reflex may be altered by suprasegmental influences on the interneuron. But the cerebral hemispheres do not work like that. When it begins to appear that a single behavioral reaction may be mediated by millions of cells, and that each cell has hundreds or even thousands of connections with its fellows, then the utility of the paradigm may be called into question. It is not just that our minds are boggled, it's our assumptions. Their limits are being stretched. We are rather in the position of the patient who knows that it hurts but can not tell the doctor exactly where. Somewhere within the conceptual framework of 20th-century neurophysiology there are some hurting assumptions which need to be identified and examined. Kuhn would probably predict that they will become obvious only when the paradigm is stretched beyond its limits.

In any case, if we could recognize and thereby analyze our assumptions, clearly we would be the better for it. An historical analysis of the philosophical origins of the paradigm could be very important in helping us overcome our intellectual obstacles to a better understanding of brain function.

NOTES

1 Edwin A. Burtt makes this point very eloquently in a slightly different context ([1], p. 225; [5], p. 273).

2 Scribner's, New York.

3 Young has traced the evolution of the associationist tradition and its influence on the development of cortical neurophysiology up to the work of Ferrier. He remarks briefly that: "The work of Ferrier was ... one of the bases of Sherrington's whole new emphasis in the study of neurophysiology, involving the use of the concepts of integration, evolution, and reflex as guiding principles" (p. 236). In his Preface and Conclusion, Young has emphasized the importance of the basic point which I am making in this essay, i.e., the generally neglected but very fundamental role of the associationist tradition in the development and practice of contemporary neurophysiology [5].

BIBLIOGRAPHY

1. Burtt, E.A.: 1932, *The Metaphysical Foundations of Modern Physical Science*, 2nd ed., Routledge and Kegan Paul, London.
2. Greenblatt, S.H.: 1974, 'Some Philosophical and Clinical Background to Sherrington's Concept of Integrative Action', *Proceedings of the XXIII Congress of the History of Medicine* (London, September 2-9, 1972), Wellcome Institute, London.
3. Kuhn, T.S.: 1970, *The Structure of Scientific Revolutions*, 2nd ed., Univ. of Chicago Press, Chicago, Ill.
4. Sherrington, C.: 1953, *Man on His Nature*, 2nd ed., Cambridge Univ. Press, Cambridge.
5. Young, R.M.: 1970, *Mind, Brain and Adaptation in the Nineteenth Century*, Clarendon Press, Oxford.

KARL H. PRIBRAM[1]

I stand accused. I have been found guilty of asserting that subjective states are coordinate with certain brain states. I also have been found guilty of asserting that there is a problem in differentiating the reference for any of these subjective states. Thus, according to my thesis, sometimes the reference is to the skin surface where energy transformations take place, sometimes the reference is introjected to bodily processes within the organism and sometimes to objects projected beyond the skin. I have therefore developed this thesis to state that it becomes useful and interesting to take the skin as an arbitrary boundary which demarcates the world within from the world out there and I have explored some of the issues and problems consequent on holding that thesis.

Marjorie Grene has taken another stance. Professor Grene suggests that we begin our explorations by envisioning the organism-with-his-world as a unit, that unit to be called the person. In my view mental states derive from brain states; in her view, as I see it, mentation is one aspect or dimension of the person. Marjorie Grene's view takes into account and emphasizes the history of organism-in-his-world to account for changes in mind or consciousness produced by brain injury; my view begins with the here and now dependence of mind on brain and brain on mind and derives history from this circularity.

Professor Dreyfus, in his accusations leveled at both Marjorie Grene and me, unfortunately misses the point. The issue is the difference in the stated assumptions that determine our views. He is therefore confused and confusing. He asserts without giving evidence that we make statements that we do not make. He muddles the term *mind* which in my presentation expressly stands for the sum of particular subjective states consensually validated Professor Grene is addressing that part of mind which lends unity to subjective awareness, i.e., self-awareness. I have elsewhere [8] addressed the problem of "Self-consciousness and Intentionality" on the basis of experimental brain research – here it suffices to make the distinction since the failure to make it has been so devastating to Professor Dreyfus' understanding.

Let me present some evidence of the confusion. Dreyfus states: "Perception of an object doesn't *seem* to be *just like* perception of an object ... that's what it *is*." I will graciously assume that his final 'it' refers to 'perception,' although this is an incorrect usage of the English language. But at least the statement is then correct, albeit trivial. If Professor Dreyfus meant what correct usage would imply, that 'it' refers to 'object,' then he is patently *wrong*. We do suffer innumerable illusions which have provided the substance of experimental psychological research for generations and patients with micropsia or macropsia or with phantom limbs attest to the occasional difference between an individual's percept and what can be consensually validated. For the patient who experiences a phantom, this difference is vital. For Dreyfus, however, "it is not clear why the phantom limb should be thought of as a projected mental state (although Descartes certainly thought of it as such), since there is no evidence that it is first felt in the mind and then projected outwards." Unfortunately for Professor Dreyfus and fortunately for patients with phantoms whom I have treated, I stand squarely with Descartes on this matter and can defend *our* position both with *proper linguistic analysis* and with *evidence*.

First, as already noted, my definition of *mind* is the congerie of subjective states coordinate with certain brain states. Dreyfus makes the statement "first felt in the mind and then projected outward." The term "in the mind" is superfluous here (what is felt is mental), but does emphasize the issue. The patient in fact "first feels the phantom" and often asks the attending nurse to massage his foot and unkink his toes *before* he is informed that these are now in the pathologist's jar upstairs. He is then acutely made aware of a disparity in his subjective state; the situation "out there" that ordinarily gave rise to his perception has changed. His perception remains essentially unchanged and for the first time he and his physician must 'really' distinguish between 'mind,' i.e., what is felt; and 'object,' i.e., what is projected. Further, clinical experience has shown that, depending on the duration of the phantom, intervention into 'mind' must become more and more central in the nervous system until only a *brain* operation will accomplish the desired change in feeling.

Dreyfus, therefore, concludes that 'Pribram is a pure Cartesian." Yes and no. I do not deny dualism but suggest how we may transcend it. In a paper entitled 'Proposal for a Structural Pragmatism' ([4], pp. 426-459), I provide a systems explanation for dualism but suggest a resolution for the problems posed by the Cartesian dilemma without sweeping the problems under a rug as Dreyfus has done here with his attempt at radical behaviorism. That Dreyfus' approach will not work can be seen in his own manuscript. First he argues that "as a philosopher one must remember that all of these investigations take place in a shared world in which we are surrounded by things and people external to us" – an assumption also shared by Pribram when he defined subjective states as involving consensual validation, a point seemingly missed by Dreyfus. But Dreyfus goes on to say that this shared world is "not in our brains or in our minds. Phenomenologically we are in a public world..." Then two sentences later he has to admit that "to be true to this phenomenon we must radically distinguish the *physical* level of interaction of external energy and internal brain states and processes, from the *phenomenological* level of persons acting in the world." *Now*, who is the Cartesian? Dreyfus claims "there is no place in this picture for a third level of mental states or processes shoved in between." But the whole point of my present argument is that had Dreyfus *read* my paper properly he would have found that his definition, and Marjorie Grene's as well, of 'persons acting in the world' and my definition of 'mental states' are operationally indistinguishable. What we apparently disagree on is the relation between 'person' or the 'mental states' that make up person and the *brain*. This is a legitimate area for disagreement but here again I claim some expertise. In still another paper, 'Toward a Neuropsychological Theory of Person' ([5], pp. 150-160), I point out once more from the standpoint of *evidence*, much of it accumulated in my own laboratory, what brain mechanisms are involved in which aspects of 'person'.

My argument thus runs: We seem all to be in agreement that (1) any analysis of the mind-brain problem must begin with the phenomenal person. Professor Grene and I both suggest, however, that (2) this phenomenal person is not a simple unanalyzable existential given, but a construction. This suggestion constitutes a paradigm change of considerable consequence which I have pursued in the structural pragmatism paper noted above. In that paper I propose (3) that analysis of person can proceed *either* to the subsystems that make up person – e.g., brain – or to the supersystems in which person is embedded. Professor Grene has, in her report to us, analyzed person in terms of his physical and social surround; I, in my report, concentrated on the person's components. Neither analysis by itself is complete. Person as a biosocial being and person composed biophysically and biochemically are complementary views of person as a whole. I have not here discussed at any length (4) the problem of person as a whole. However, Professor Grene did tackle this question using the data obtained from patients with callosally split brains. As noted earlier, I have attempted elsewhere to handle this problem by differentiating self-consciousness from the consciousness of perceptual awareness. Further, I have pointed out that psychologists since William James have experimentally approached this issue in terms of the direction and span of attention.

In closing, I want now to illustrate with an example the utility of the component analysis in clarifying a specific set of problems raised in this conference. We heard earlier of the dilemma posed by masochism for an operational analysis of the problem of pain. We heard further that suffering appears to have a voluntary component. Thus pain appeared, on occasion, to be pleasurable and suffering a result of willed choice. Though consonant with the known facts, these observations appeared to be sufficiently paradoxical to bring into question the sanity of observed and observer alike.

An understanding of the physiological mechanisms of pain – the brain part of the mind-brain whole – clarifies the apparent paradox in a way that attention to mind (or person), alone – the exercises we heard presented – cannot. The problem was given us in two parts: (1) pain *vs.* pleasure, and (2) minding pain; the analysis proceeds accordingly.

First, *two* neural systems deal with the effects of nociceptive stimulation. One system is composed of fairly large peripheral nerve fibers that convey impulses rapidly to the brain stem, and thence to the parietal cortex. This fast system has been known for some time to deal with locating the nociceptive stimulus and was termed *epicritic* by Henry Head [2]. Another system, much more interesting for our present purposes, utilizes very fine peripheral nerve fibers to more slowly transmit the effect of nociceptive stimulation to the brain. The terminations of these fibers and the functional mechanisms of the brain parts in which this slow pain system terminates are only now beginning to become clear. I have recently suggested the term *protocritic* for this system since it apparently deals with the elementary intensive aspects, not only of nociceptive, but of all sensory stimulation.

A striking discovery has just been made regarding the protocritic system. We heard earlier of the gate theory of pain proposed by Melzack and Wall [3]. This theory has the advantage that it explains the fact that under certain circumstances nociceptive stimulation fails to produce pain which on other occasions would result from identical stimulation. The recent discovery is that in addition to neural gating, a substance is secreted that acts as a chemical gate. This substance has protective properties against pain, similar to those of morphine and is thus labeled MLS – morphine-like-substance. This discovery of a pain protective substance places the pain mechanism along with others which depend on the chemical sensitivities of the central nervous system. (The central core of the brain stem is endowed with a variety of receptors sensitive to blood sugar level, osmolarity, partial pressure of CO_2, catechol and indole amines, to name just a few of the most potent of these chemicals monitored by the brain stem structures.)

Characteristic of these chemically sensitive mechanisms is their homeostatic organization. Control over the concentration of the substance to which they are sensitive is obtained by means of a feedback operation. A gate is such a feedback. The control of pain, therefore, is to be conceived in terms of homeostasis.

Homeostatic mechanisms display an appetitive and a consummatory phase. The pain homeostatic mechanism apparently is no exception. As we heard earlier, itch, the masochistic ritual that serves as a prelude to orgasm, and even the initial orgastic buildup, are appetitive in their manifestations. And, as in all appetitive processes, whether they are perceived as pleasurable or painful depends on a variety of intensive factors such as incrementation, past history (expectation), and duration.

But this is not all. Professor Bakan, in his superb presentation, addressed the issue of minding pain as in part a problem of volition. The feedback homeostatic pain mechanisms of the brainstem, described above, have little to offer by way of explaining this second paradox. But as Professor Bakan noted, pain must often be suffered, and frontal lobotomy has been found to relieve suffering. Suffering comes about when the ordinarily homeostatic *negative* feedback of the pain mechanism is converted by faulty timing, failure to consummate, etc., into a positive regenerative feedback which produces accruing oscillation rather than stability. Such a positive feedback can be brought under control by the intervention of an override [1], a process that converts the feedback to a feedforward. Feedforward processes are characterized by a parallel rather than a hierarchical organization ([6], chapter 5). Minding – i.e., paying attention (as defined by Ryle and used in my earlier presentation) – initiates such control on the homeostatic feedback operation of the pain system by way of providing parallel simultaneous sensitivities much as do the currently popular biofeedback procedures [8]. In minding, whether by way of biofeedback or more ordinary means, previously automatic processes are brought under intentional, voluntary control, much as an automatic thermostatic feedback mechanism is converted into a more flexible system by the introduction of a bias, the little wheel on top of the instrument by which one can alter the set point (the temperature) around which the feedback will become stabilized. I have elsewhere reviewed the large body of experimental evidence which indicates that the frontal cortex of the brain operates as an executive to the rest of the brain, ensuring flexibility of operation by a mechanism similar to the introduction of a bias on a thermostat ([7], pp.293-314).

This, therefore, is an example of the utility of a brain functional approach to problems posed by philosophical analysis. The paradox of masochism can be understood when it is realized that the mechanism of pain perception involves a homeostatic feedback process which has an appetitive and consummatory phase. The paradox of the voluntariness of suffering can be understood when it is realized that feedforward intentional operations as well as feedbacks are involved in the control of pain.

The philosopher may argue that such scientific understanding is incomplete, and his argument must be honored. Scientific understanding is never complete. There remains an artful mystery to the proper production of a symphony by a piece of corrugated cardboard, moved by a magnet, controlled by a stereo high fidelity system even after we scientifically know the characteristics of each of the components and have complete access to circuit diagrams and the like. Still, scientific understanding is enriching. In the analysis of the mind-brain problem, the topic of this conference, neuroscientific understanding can contribute enormously. To paraphrase Marjorie Grene's brilliant closing comment at our session: the recent, often astounding, discoveries in brain function have, on occasion, made the neo-Cartesian scientist seem out of this world. This, however, is perhaps to be preferred to the radical behaviorism which, stemming from a positivist tradition, has driven the philosopher out of his mind.

NOTES

[1] This work was supported by NIMH Grant No. MH12970 and NIMH Career Award No. MH 15214 to the author.

BIBLIOGRAPHY

1. Ashby, W.R.: 1960, *Design for a Brain: The Origin of Adaptive Behavior*, 2nd ed., John Wiley & Sons, New York.

2. Head, H.: 1920, *Studies in Neurology*, (two volumes), Oxford University Press, London.
3. Melzack, R., and P.D. Wall: 1965, 'Pain Mechanisms: A New Theory', *Science* **150**, 971-979.
4. Pribram, K.H.: 1965, 'Proposal for a Structural Pragmatism: Some Neuropsychological Considerations of Problems in Philosophy', in B. Wolman and E. Nagle (eds.), *Scientific Psychology: Principles and Approaches*, Basic Books, New York,
5. Pribram, K.H.: 1968, 'Toward a Neuropsychological Theory of Person', in E. Norbert *et al.* (eds.), *The Study of Personality: An Interdisciplinary Approach*, Holt, Rinehart and Winston, New York.
6. Pribram, K.H.: 1971 *Languages of the Brain: Experimental Paradoxes and Principles in Neuropsychology*, Prentice-Hall, Englewood Cliffs, New Jersey.
7. Pribram, K.H.: 1973, 'The Primate Frontal Cortex – Executive of the Brain', in K.H. Pribram and A.R. Luria (eds.), *Psychophysiology of the Frontal Lobes*, Academic Press, Inc., New York.
8. Pribram, K.H.: (in press), 'Self-Consciousness and Intentionality', in G.E. Schwartz and D. Shapiro (eds.), *Consciousness and Self–Regulation: Advances in Research*, Plenum Publishing Corporation, New York.

ROBERT M. VEATCH

Asking the function of philosophical concepts is rather like asking the function of matter or energy or nature. On the one hand the functions are infinite. In terms of classical functional analysis, philosophical concepts are symbols expressing patterns of all conceivable functions: patterns of basic meanings and systems of belief; integrative world views, norms and values; goals to be attained; and means for adaptation.[1] In the neurosciences, in the sciences more generally, and in all organized and symbolic forms of interaction, philosophical concepts serve all functions which can be symbolized in terms of philosophical categories.

On the other hand there is something wrong, even offensive, in the question, "What is the function of philosophical concepts in the neurosciences?" It can be offensive to individuals in their roles as rational people and neuroscientists for two very different reasons. To reasonable people, particularly philosophers who may occasionally be included in that class, it is objectionable to ask the function of philosophical concepts because to ask the function of something is to imply it serves some more ultimate purpose – that it is instrumental and to be analyzed and explained in terms of the purpose it serves. In other words, if philosophical concepts cannot be defended in terms of their payoff for the neurosciences, they are "useless" or not worthy of our attention. It could, of course, be that analysis of philosophical concepts could be useful to the neurosciences. But philosophical concepts must also have intrinsic as well as instrumental or functional values. To the person of reason, philosophical concepts are not to be analyzed simply in terms of their function, but rather also in terms of their structure or the substantive legitimacy of their content.

The neuroscientists, on the other hand, might be puzzled by the question of the function of philosophical concepts for quite different reasons. The neuroscientists, whom I knew while working in neuropharmacology at the University of California Medical Center at San Francisco, were pragmatic people. That philosophical concepts had anything to do at all with the pragmatic solving of the scientific puzzles that occupied their day–to–day life would come as a great surprise.

The analysis of the function of philosophical concepts in the neurosciences being my perplexing task, I would like to propose a rather simple, two-edged thesis: that philosophical concepts are thought by many – including many neuroscientists – to be irrelevant to the neurosciences, and yet logically they must dominate the thought structures of both the neuroscientist and those patients and professionals who, at the clinical level, make judgments which apply to results of the neurosciences to clinical medicine. I would like to defend (or illustrate) this thesis with two personal examples. The first illustrates the pervasive impact of philosophical concepts in a neuropharmacology research laboratory; the second, their impact on clinical decisions from the days of my life as a medical ethicist in a major teaching hospital. Both are illustrations drawing on a concept fundamental to both the neurosciences and philosophy: the concept of pain.

1. *Pain in Neuropharmacological Research*

When I went to do graduate work in neuropharmacology I went with a rather strange question: Why is it that pain hurts? I asked the question as a social reformer who disliked pain and suffering and would have liked to have done something about it. I also asked the question as one puzzled by the concept of pain. Pain seems like a physiological sensory phenomenon, pure and simple. We understand the nerve pathways from pain receptor into the spinal cord and into the brain stem. At least in my days of neurophysiology the somatic pain pathway passed from receptor into dorsal root of the cord, into the posterolateral ventral nucleus of the thalamus and, finally, to the somesthetic area of the cerebral cortex. The pharmacology of pain was the simple process of blocking electrochemical information flow.

My question, however, was a different one. How is it that this pure and simple electrochemical process – this flowing of sodium and potassium ions which generated an electrical charge and in turn a nerve impulse – could give rise to a totally different type of response – a response which says "I do not like that." It is a violation of the laws of conservation of matter and energy. We begin with sodium and potassium ions, with electrical charges and nuclear particles, and end with something that can only be identified as value – a judgment that that particular flowing of charged particles is clearly and decisively evil. Somehow in the process of electrochemical transmission from peripheral receptor to central brain tissues there has been a fundamental transition from something that sounds like physics and biology and chemistry to something that sounds like philosophy or psychology or value theory. "I do not like that" is not supposed to be the matter of the lab sciences. Yet, "I do not like that" is essential to our concept of pain, and pain is central to the branch of neuropharmacology I chose to study.

To make matters more confusing, philosophy or religion or the realm of values is normally thought of as soft or subjective or purely relative to one's personal feelings, and yet the pain resulting from, say, sticking a pin in my finger gives rise to an "I do not like that," which has very objective qualities. If we found a patient in neurology in whom we could stick a pin and not get the "I do not like that" response, we were convinced we should try to find out what was wrong. While values in general are relegated to the realm of tastes or subjective feelings, pain is remarkably uniform and attributed, to use the gestalt term, with objective localization in spite of the fact that it is clearly an evaluative phenomenon – a value judgment that an experience is bad.

I pushed my problem with my pharmacological colleagues with little success. When I asked why pain hurts, I would be told of the neural pathway, of the unique qualities of a pain receptor, and of the hypothetical receptor sites in the brain where the pain phenomenon was recorded. If pressed, my colleagues would confess there was much we did not know about pain perception; that once the nerve impulse got to the thalamus, neurophysiologists seemed to lose track of it, but it was always with the tone that someday there will be a breakthrough and we will solve the problem.

We have seen the same kind of anatomical answers to the question in the recent Western attempts to rationalize acupuncture. Why does acupuncture work as an anesthetic? We are told that impulses generated by the pin interfere – perhaps in the diffuse thalamic projection system – with the normal transmission of pain. There are recent references to "gate-theories" of pain in the literature. These anatomical explanations of pain, however, simply describe how impulse transmission might be interrupted.

But my question was a different one; not how can we further understand the neural pathway for pain transmission but, more fundamentally, why is it that when those pathways are not interrupted the electrochemical complex can end up hurting? That was a question my pharmacological colleagues could not handle. It was as if I had violated the fundamental laws of the neurosciences. There is a blood-brain barrier which is very fashionable in neuropharmacology. It seems there is also another barrier: the mind/body barrier. It is obvious that a very basic phenomenon, the phenomenon of pain, crosses that barrier; yet as a neuroscientist I was not allowed to cross it. The radical dichotomy between the scientific and the evaluative – between the mind and the body – could not be bridged, at least with the conceptual tools available to me in the neurosciences of the day.

Even more disturbing is the fact that other disciplines – including the disciplines of the humanities – do not feel comfortable with the question of why pain hurts either. Philosophy and theology appear not to be able to bridge the mind/body dichotomy. Theology gives us the theodicies such as Job. Philosophy gives us cosmologies of suffering. Both attempt to account for why it could be that suffering is included in the cosmic order. Psychology can account for why specific experiences hurt and even why it is that some individuals who apparently receive the pain nerve impulses claim they do not mind or even enjoy the pain. None of these disciplines, however, seems comfortable with the question of why an electrochemical process can end up hurting in the first place.

2. *Pain in Clinical Medicine*

My struggle to bridge that unbridgeable barrier led me to the field of medical ethics and to my second illustration of the impact of philosophical concepts on pain. Seeing patients at a major teaching hospital, I encountered a man in excruciating pain from a compound leg fracture. In semi-coherent fashion he was pleading with the nurse, with me, and with anyone who would listen for something to relieve his pain. Soon after, I met the resident responsible for the case and asked what medication had been given for his pain relief. The response was Darvon Compound 65 mg – a non-narcotic analgesic claimed to be more potent than aspirin, but less than the weakest narcotics. I asked why the man was not given something stronger, some Demerol perhaps. The physician's response was that he felt he should not take a chance with the narcotic. He mentioned in rather disorganized fashion the risks of respiratory depression and the danger of creating an addict, especially when the pain would subside naturally in a day or two. The physician's task, according to this young practitioner, was to intervene very cautiously and let nature take its course.

I was able to bring another physician into the discussion, one whom I suspected would present a different so-called medical evaluation. He was indeed greatly disturbed by the case. He thought that it was outrageous that a patient should suffer when man, through his technological genius, had created drugs which would conquer that evil of pain. Man's task, according to this physician's response, was to use his ingenuity to conquer the evils of nature.

The philosophical concepts and the systems of belief and value are perhaps obvious. What was striking, however, was that neither physician realized that he was operating with his own particular set of philosophical concepts. The first physician operated with a theory of nature and its inscrutable mysteries, of the fallibility of man when he tries to tamper with nature, and of minimal evil of suffering when compared with other risks of iatrogenic respiratory depression and addiction. He had inherited a concept of physiological homeostasis from Walter B. Cannon and indirectly from those who have a static view of the world. Homeostasis, which was taught as a physiological fact in my physiology courses, is a dangerously conservative philosophical concept – but for our purposes the point is that it is a philosophical concept. It is transmitted into clinical neurology in its concept of the normal and the imperative to return the patient to normal which takes precedence over other possible uses of the neurosciences such as improving on nature's creation.

The second physician also acted on a set of philosophical concepts.

Nature was something to be conquered; suffering was something to be overcome. If pushed, he probably would not have been as comfortable with the conservative homeostasis concepts as with the more dynamic world views of a Bergson or a Darwin or a Marx who saw movement or evolution or change much more of the essence.

Both of these physicians, however, shared the belief in the mind/body dualism. Both believed they were engaged in what they would see as good scientific medicine. To have suggested that they were engaged in a struggle between competing philosophical concepts would have appalled them. Each would have felt uncomfortable treating the patient's suffering if he thought it had its origins in the mind or in a system of beliefs and values. Why is it that the neurologists strive to find an organic basis for the pathological behavior before they intervene with drugs or surgery? The comforting Cartesianism which was first encountered in the pharmacology lab is also at home in the clinical setting.

Yes, philosophical concepts do function in the basic neurosciences and the clinical decisions alike. They function in all the possible ways that concepts can function: to provide a basic framework of understanding, to symbolize goals, to articulate means for adapting, and to frame norms and values. They function to make certain questions – like why pain hurts – unaskable, while making other questions very askable. They also function to insulate the neuroscientist from an awareness of the basic commitments to beliefs and values which permit him to act. Perhaps this hiding of philosophical concepts and their function is essential to the smooth functioning of neuroscientists. Perhaps they would be paralyzed if they examined the life which they lead. But in the end it may be that the functions of these philosophical concepts cannot remain hidden. The neuroscientists who pioneer, the ones who really understand and make breakthroughs, will be the ones who realize that philosophical concepts are essential to their task, essential not because they serve particular functions, but because it is logically necessary to use them to carry on the human enterprise.

NOTES

[1] Classical functional analysis to which I refer is exemplified by sociologists of culture including Max Weber, Emile Durkheim, and especially for the 20th century, Talcott Parsons. See his 'Pattern Variables Revisited: A Response to Robert Dubin', in 1967, *Sociological Theory and Modern Society*, The Free Press, New York, pp. 192-219.

IAN R. LAWSON

Drs. Greenblatt and Veatch have remarked that we clinicians rely a lot more on assumptions than we recognize. May I say additionally that some of these assumptions are not benign, are not simply "gaps" in knowledge, but constitute aphorisms and premises (frequently overlaid by sentimentality or excuses of "pragmatism") that obstruct us from better thinking and ultimately from better patient care. At any rate, all such should be dissected utterly and without pity. Such is the complexity of the issues that we physicians are incapable of doing this unaided. Welcome, then, to the philosophers!

In my own case, I supervise a small part of the one million long-term care beds devoted in America to the care of the disabled and elderly, many of whom are neurologically grossly disabled. But I spend much of my professional time attending not to their clinical needs but to the dysfunctional cybernetics that the able-minded have created around the care of such disabled: created on the basis of such unexplored axioms as 'pluralism,' 'voluntarism,' 'pragmatism,' and the 'free enterprise system'. An unsparing dissection of these could be a most compassionate activity.

So much for the health care context. There are also many clinical situations where the sick could be better served by clear, particularised thinking rather than by aphorism-laden concerns that stop short of it. Especially towards the brain damaged and neurologically impaired, I observe that our intellects function adequately in the initial diagnostic effort but lapse when it comes to the management of those for whom specific 'cure' or treatment is still lacking. Walther Riese makes this point in the last chapter of his little history of neurology. It is an attitude that could be summarized by "You have multiple sclerosis. My professional capacity is limited to the labeling. I am awfully sorry. Now you need a friend."

Yet, in the past, neurologists frequently undertook a more continuous care of the neurologically disabled as part of their discipline. The foundations of scientific neurology were laid by such as Charcot who were involved in institutional long-term care. Were they custodians to museums of which modern medicine no longer has need? Certainly, when it comes to consideration of mind-body relationship, those specialties that have maintained more continuous care of the mind and the body, that have pursued that responsibility longitudinally through cycles and phases to life's end, such may have a better experience from which to speak than those who selectively maintain a limited, early, and largely diagnostic contact. The specialties, for instance, of renal dialysis, medical rehabilitation, and geriatric medicine spend a large part of their day-to-day efforts considering mind-body interrelations as matters of practical effect, not as a theoretic add-on to a structural interest in brain and nervous system lesions. As an internist, one observes fluctuations of mental with the metabolic state, *le milieu intérieur*, which often provide vivid and different insights from those obtained from observation confined to the nervous system alone.

I now present a case for clinical-philosophical discussion. You are all aware of the importance that the clinical-pathological conference has in medical education. The case is recounted in its clinical history and features, then the guest speaker rises to risk his reputation on the diagnostic hypotheses that he develops, and at the end the pathologist's report vindicates him or otherwise. Lawrence Weed has recommended an entirely different approach, a problem-oriented one, which would allow a logical review of the serial decisions that were made on the basis of the knowledge that was available at the time each decision was made. But it is hard to eschew the chance of a fine display of brinkmanship, so the conventional conferences persist. Perhaps this indicates that not only do we clinicians rely on assumptive judgments – we love 'em!

As a clinician, I may say that I have found the involvement of a philosopher, Stuart Spicker, stimulating and fruitful in both teaching and research. Case conferencing played an important part in a medical student elective seminar organised by Stanley Ingman, our colleague in medical sociology. My case today for clinical-philosophic conference is designed to raise questions and propositions from known outcomes.

At the age of 63 years, this man sufferd a calamitous right-side stroke. After a few weeks in a general-medical service, he was transferred to my geriatric assessment unit in the Aberdeen Teaching Hospitals, where he spent one year and 4 months before discharge home.

In 1952, at the age of 50, he felt peculiar axial pains in the anterior chest and rectum which were accompanied by minimal left-sided pyramidal signs. Four years later, he felt regularly "dizzy and hazy, and goes on his knees on the mat," such attacks lasting ten minutes at a time. He was observed to mumble his words. He also suffered from angina pectoris. Ten years later, in 1962, he began to take aspirin regularly for back pain, and in 1963 was admitted to hospital with rectal bleeding, a recognized complication of his own treatment. Three hours after admission, he became emotional and tremulous. Bilateral pyramidal signs were observed, which progressed on the right side to frank hemiplegia with hemi-anesthesia. Such was his described state on transfer to my care several weeks later.

Observation over the ensuing months yielded the following detailed picture. Not only was he mentally alert and fully oriented, but he retained an enormous drive to achieve optimal recovery. He wept easily when frustrated or when topics of sentimental association were mentioned. He had stuttering and word-finding difficulty which did not prevent him from giving an insightful account of his disability.

Subjectively, he described four unpleasant forms of sensation. Least troublesome were episodic visual hallucinations ("coils," "lights," "colors," accompanied by "muzziness") and noises in the ears ("seashells"). Very distressful was the constant sense of "an abyss" on his affected right side, into which he felt that he was about to topple. (I have had several patients complain of this who had severe extra- and proprio-ceptive losses from right cerebral damage.) Paroxysmal, intense, occipital headaches were his fourth sensation.

He had a full right homonymous hemianopsia to finger movement and a constricted left visual field was identified on Bjerrum screen testing. Functionally, he often appeared to be reduced to tunnel vision. He read printed texts with difficulty, and he was aided by slow lateral oscillatory movements of the text. He had gross impairment of conjugate deviation of eyes in all directions, both to command, attraction, and on following objects. Both right upper and lower limbs were completely powerless but, in addition, his left upper limb was somewhat weak and his left lower limb conspicuously so. He had complete cutaneous anesthesia down his right side. It also involved the posterior aspect of his left leg. Two point discrimination and stereognostic discrimination was also impaired in his left fingers. There was profound proprioceptive loss in his right upper limb, which he called his "dummy". When his eyes were closed, he failed to locate his right thumb with his left hand and he located (or preferred?) my thumb in lieu (Figure 1). There was loss of joint sense, not only on his right side but in his left lower extremity as well. He had no rectal sensation to the insertion of a gloved finger but felt quite acutely a small peri-anal pustule

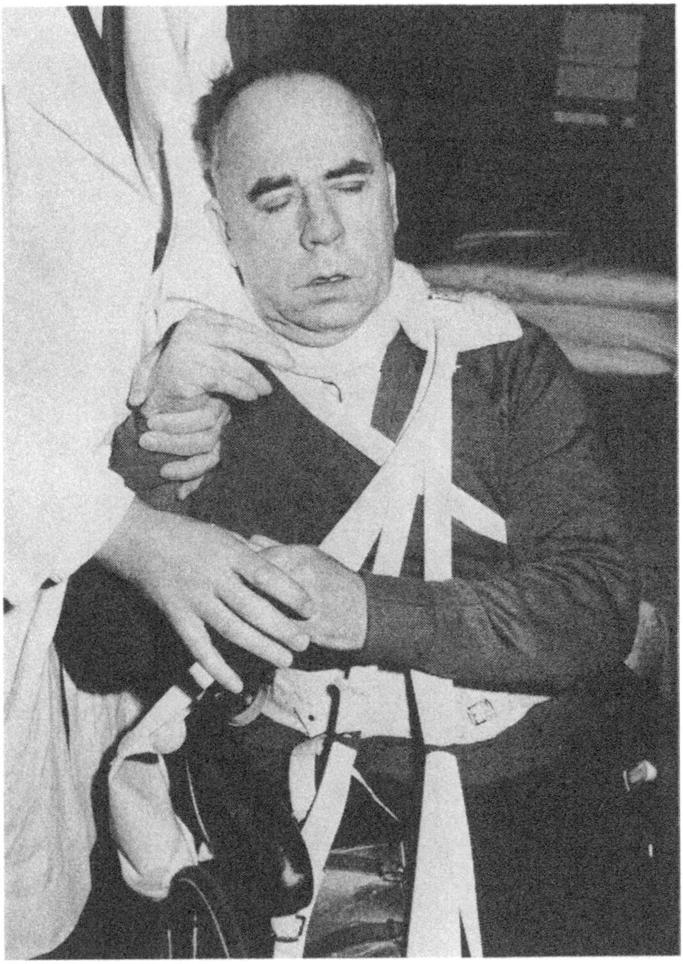

Fig. 1. This identifying photograph is included with the express permission of the patient.

that developed later. He had little awareness of bowel and bladder sensation and maintained cleanliness by meticulous, routinised visits to the toilet.

Overall, it seemed likely that "four vessel" arterial disease in the neck was responsible for his cerebral ischemic symptoms, with early symptomato-

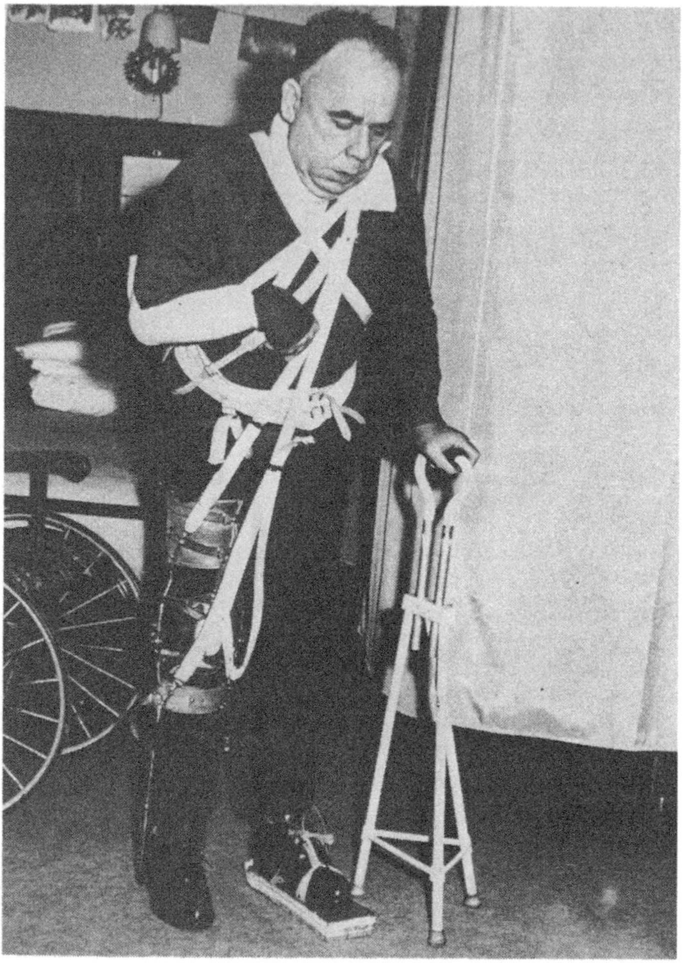

Fig. 2. This identifying photograph is included with the express permission of the patient.

logy in the vertebral-basilar distribution. Arch aortography was unsuccessful, however, causing a large hematoma and unwellness.

Responding to his determination, a comprehensive rehabilitation program was deployed involving occupational, physical, and speech therapists, and a skillful, senior male nurse. There were recurrent fluctuations of physical function and well-being attended by headaches, "muzziness," dizziness, and deterioration of vision. Imipramine alleviated the emotionalism, analgesics the headaches, and a supportive cervical collar to the neck helped the dizziness. For walking (Figure 2) a complicated apparatus was developed by the physical therapist (Miss Ormiston) and the nurse (Mr. Bell), the essentials of which were the bracing of the paralyzed right leg (to maintain extension), elevation of the left side by a block under the left shoe (to aid swing-through of the right leg), and a harness from the left shoulder to the right leg (by means of which he moved and sensed that limb). A sling held his paralyzed right arm.

The patient, Mr. C., describes his walking thus:

Miss Ormiston learned me to hold my balance on the parallel bars. I kept counting and feeling the tension with my left hand. Now I carry it out with tripod in my walking. I first take my balance by the certain tension on my left arm, then at the count of one I move tripod forward. No. 2, I move left foot forward. Then with count of three, with left shoulder I can lift my right leg along. Sometimes Mr. Bell ... makes adjustments to right arm which somehow effects my balance. I tried out walking on the carpet in the room of physio-therapy and walked 3-4 yards – much more difficult than walking in the corridor (on a smooth surface). It takes me a minute or so to think what I am going to do next. After walking so far, I get tired and my left leg gets weak and refuses to take a grip.

This 'walking' was so manifestly exacting that I questioned its effect on his damaged heart, the more so when electrocardiography showed exercise depression of his ST waves with tachycardia. However, an energy output study by an obliging colleague of the M.R.C. Obstetric Medicine Unit (Dr. F.E. Hytten) showed that in 15 minutes of such walking, the energy cost (3.1 Kcal. per minute) was little more than the cost of normal walking over the same period. In that time, however, Mr. C. had walked only 30 yards, 2 yards per minute. Hence, for practical mobility he propelled himself in a wheel chair with one-arm drive, 'walking' being done twice a day at his own insistence. Over the months, he attained sufficient independency to be discharged home, returning for day-hospital care regularly thereafter. He was still active in this regime on my last visit to Scotland in 1973, ten years after the massive stroke.

From this case I would put forward the following propositions for discussion:

(1) Brain damage, like all disability, makes daily living a process of thinking and mensuration. A line of sequentially metered, conscious acts

now replaces a range of freedom unconsciously engaged in 'at will'. Brain damage may indeed result in an over-dependence on the logical, manifest as 'concretism'. While this is useful and inescapable for certain functions, brain damaged patients may have to be released from total limitation to such logic-bound thinking to re-engage in the perceptual risk-taking and errors characteristic of 'normal' life.

(2) Disability involves a thinking, circumspect regard of the minutiae of environmental circumstances – such as the texture of the floor, width of a door, height of a toilet seat. In this respect, the physical environment is inspected from the aspect of a mountain climber. It is no longer the passive, supportive substrate on which normal life proceeds. It is intrusive and pre-occupying.

(3) Successful use of an external prosthesis involves a complex relation to the body image, especially when attached to a lost half-side. While prostheses always carry a traumatic potential (in this man's case, the pressure of the left shoulder harness produced dizziness, probably from pressure on the origin of the vertebral artery on that side of the neck), nevertheless rejection or assimilation of the prosthesis into the body image is not accountable to that alone. Note too, that in this man's case, the harness was as much a sensory prosthesis as a motor prosthesis, aiding feedback to the intact left side. Nevertheless, it did not reconstitute the body image as an artificial limb does by filling in for the lost part, whose image is retained by an intact brain. For what has been lost is not the part but sense of it, lost or present.

(4) The mind of Mr. C. maintained an integrity of logic, feeling, motivation and insight; but also included the effect of its own brain destruction and associated losses felt as a novel, positive symptom – 'the abyss.' Around half a body image, he was able to reconstruct a function normally dependent on the whole. Is the price or index of usability of a residual body image an awareness of how partial it is?

(5) Disease and disability in such a case cannot be considered as 'health minus whatever' or 'normality minus whatever.' It is a characteristically unique experience. It is profoundly different from normal living. Into it the clinician must enter in a kind of cerebral symbiosis with his brain-damaged patient, learning most from the ingenuity with which a residual brain circumvents and substitutes for its own disability.

MARJORIE GRENE

CHAIRWOMAN'S CLOSING REMARKS

Our panelists have illustrated dramatically the conceptual complexities and at the same time the rich promise for philosophical reflection as well as for medical actuality that meet us at the interface of what is sometimes called the philosophy of mind (what I prefer to call the philosophy of the person) and the principles and practice of the neuro-medical sciences.

First, the complexities. As so often happens, it is the historian who comes to remind both philosophers and scientists of the overabstractness of our approach to the problem we have all been talking about: the problem of mind and brain: of the experienced 'consequences' (?) or 'side-effects' (?) or 'achievements' (?) of the immensely complex and subtle neural structures and functions on which, for our existence as human beings, we certainly depend. We usually – both philosophers and neurologists – worry about something we call 'Cartesian dualism,' whether to attack or to defend it. Lashley, for example, in the paper Dr. Pribram and I cited in our papers, wants to attack it; Sherrington in his philosophical work, so strangely at odds, as Dr. Greenblatt points out, with his early reliance on the associationist model for the explanation of behavior, wants to defend it – even in his classical *Integration of the Nervous System*, which manages, as Dr. Greenblatt suggested, to be both S-R oriented and openly dualistic. Yet the recognition of our kinship with other animals – and surely we do all recognize that kinship – *should* produce in us an opposition to Cartesianism, not on behalf of the abolition of mind in favor of 'mere' unfeeling machinery (let alone uninhabited public institutions), but on behalf of the sentience, the 'centricity' (to borrow my rendering of Adolph Portmann's concept *'Weltbeziehung durch Innerlichkeit'*), the lived-bodily existence in a characteristic real environment, both inner and outer, both biological and physical, of *all* living things. Dr. Bynum describes this attitude as 'vitalism,' an unfashionable word, yet one which, in the way he uses it, provides a salutary reminder of the fact that both mind and body (whatever those difficult concepts mean) are grounded in the phenomenon of life. It is *not* a question nowadays, as Professor Dreyfus in his earlier comments seemed to think, of the choice of a two-substance theory, such as

Sherrington came to embrace, and a one-substance theory, which views matter-in-motion, including the bête-machine, as all there is. Even though their matter is the same, living things are organized differently from non-living things: witness the fact that they are all encoded, all professionally, hereditarily purveyors of information. Further, all living things with central nervous systems are differently organized, again, from living things without such systems, in that, to a greater or less degree, they are centers of action within the environment appropriate to their species. They receive and furnish information, not only through their hereditary material, but in their experience of the environment and their action on it. Each has, as von Uexküll put it, its *Merkwelt* and its *Wirkwelt* – not that these are independent of one another either; they are deeply interdependent (just think of Held's kittens). And finally, living things with central nervous systems like ours act in and on, not only their biological environments, but also the cultural world(s) through and in which they achieve human maturity, in a way, so far as we can tell, unique even among 'higher' (or recent) vertebrates. This is *not* a question of 'a' mind, distinct from, or identical with, or parallel to, a body, let alone of public institutions inhabited by no one, but of *our* bodily-mental, neurologically-embodied, inwardly-lived style of animal existence, as distinct from the myriad life styles of bees or dolphins, of tree mice or sticklebacks, of chrysanthemums or spirochetes. And, of course, Dr. Pribram is quite right in insisting on how much the neuroscientist can teach philosophers about the necessary biological conditions for this peculiar life style. The context in which Dr. Bynum has put the problem, though, should help us beware of too easy bandying of the concepts 'mind' and 'body', however, as philosophers *or* neurologists, we wish to relate them. Mind is how we live our bodies. With a 'normal' brain we live them routinely and relatively unthinkingly; if we live with an impaired brain, like Mr. C.'s, the whole business, as Dr. Lawson has so brilliantly shown us, takes on so different a cast as to acquire a different physiognomy, a different 'essence' from that of an ordinary life. But what is involved is always style of *life*, not just of being, whether intellectual or physical.

There are other complexities which seem to me sometimes in danger of becoming confusions. If it is not just a question of mind/body we are considering, the question of mind/body itself is not identical either, as Dr. Veatch seems to suggest, with the contrast: neuroscientific/philosophical. Neuroscientists think, too, even when not philosophizing;

Mr. C. thought very hard indeed in order to master the skill of locomotion in reliance on his grossly impaired brain and heart. But neither the practicing neurologist nor Mr. C. is *philosophizing*. Philosophy is thought reflecting on itself, an activity in which, as both Dr. Greenblatt and Dr. Veatch suggested, Dr. Neuro at work has neither time nor inclination to indulge. We must recognize this distinction, it seems to me, precisely in order to be able to draw out, as, again, Drs. Greenblatt and Veatch rightly want us to do, the unrecognized assumptions of working neuroscientists which constitute, in a sense, their 'philosophy': the fundamental beliefs on which their everyday reasoning, and, ultimately, their fateful day-to-day decisions, do indeed depend. But to call all thought or all appraisal or all evaluation 'philosophy' is to miss the subtleties of the very intellectual situation we want to analyze and understand and perhaps improve. (Not, please, Dr. Bynum, that philosophy is always, or even often, prescriptive: it analyzes, but, for the most part, neither proposes nor disposes. If we can help physicians to reflect more clearly on their basic principles, *they* will have to decide what to do with their new enlightenment, much as we have to learn to reflect differently in the light of their new knowledge.)

That brings me to the promise. The network of problems in which, in the past few days, we have been entangled, from the puzzles of pain, whether hurtful or, sometimes and paradoxically, attractive, to the ethical challenge posed by psychosurgery: this nest of problems has led, in part, of course, to misunderstandings or just missing of minds; it has certainly not produced a finished pattern of thought for neuroscientist and philosopher to follow. Yet how varied, how full of content, how tantalizing are the horizons opened up by this closing session alone: the perspectives of medical and biological history, the possibility of uncovering the hidden assumptions which may influence the development of research and practice as well as the direction of philosophical reflection, the lessons advancing research in neurophysiology may have for the philosophical problem of mind or person: all these are so many open doors for further conversation and research. Most of all, for me at least, Mr. C. bears heartening witness to the fruitfulness of the concept of the lived body, of the person understood as relying on the natural instrumentalities of his bodily being *and* on the artificial instruments by which he supplements them in order to focus on a task he is determined (in the voluntaristic, not the causal sense) to perform. Dr. Lawson's superb account of what it is to be diseased shows up by contrast the very different structure of what it is to be healthy, and in particular it demonstrates the fruitfulness of the concept of the body schema, the centrality of the notion of the lived body for the philosophical as well as the scientific understanding of human nervous systems, and of the human beings who possess them, in sickness or in health. May we meet again soon, for there is much that we have to learn from one another.

NOTES ON CONTRIBUTORS

David Bakan, Ph.D., is Professor of Psychology, York University, Downsview, Ontario, Canada.

Arthur L. Benton, Ph.D., is Professor of Neurology and Psychology, The University of Iowa, Iowa City, Iowa.

William F. Bynum, M.D., Ph.D., is Lecturer in the History of Medicine, University College London, London, England.

Hubert Dreyfus, Ph.D., is Professor of Philosophy, The University of California at Berkeley, Berkeley, California.

H. Tristram Engelhardt, Jr., Ph.D., M.D., is Associate Professor of the Philosophy of Medicine, Institute for the Medical Humanities, and Associate Professor in the Department of Preventive Medicine and Community Health, The University of Texas Medical Branch, Galveston, Texas.

Jerry A. Fodor, Ph.D., is Professor of Philosophy and Psychology, Massachusetts Institute of Technology, Cambridge, Massachusetts.

Samuel H. Greenblatt, M.D., is an Instructor in Neurological Surgery, Albert Einstein College of Medicine, Yeshiva University, New York City, New York.

Marjorie Grene, Ph.D., is Professor of Philosophy, The University of California at Davis, Davis, California.

Hans Jonas, Ph.D., is Alvin Johnson Professor of Philosophy, The Graduate Faculty of Political and Social Science, New School for Social Research, New York City, New York.

Ian R. Lawson, M.D., F.R.C.P., Edin., is Medical Director, The Hebrew Home for the Aged, and Associate Professor of Medicine, School of Medicine, The University of Connecticut Health Center, Farmington, Connecticut.

Joseph Margolis, Ph.D., is Professor of Philosophy, Temple University, Philadelphia, Pennsylvania.

George Pitcher, Ph.D., is Professor of Philosophy, Princeton University, Princeton, New Jersey.

S. F. Spicker and H. T. Engelhardt, Jr. (eds.), Philosophical Dimensions of the Neuro-Medical Sciences, 269–270.

Karl H. Pribram, M.D., is Professor of Neuroscience, Stanford University, Stanford, California.

Jerome Shaffer, Ph.D., is Professor of Philosophy, The University of Connecticut, Storrs, Connecticut.

Stuart F. Spicker, Ph.D., is Associate Professor of Community Medicine and Health Care (Philosophy), School of Medicine, The University of Connecticut Health Center, Farmington, Connecticut.

Bernard Tursky is Professor of Political Science, and Director of the Laboratory for Behavioral Research, The State University of New York at Stony Brook, Stony Brook, New York.

Robert M. Veatch, Ph.D., is Associate for Medical Ethics, Institute of Society, Ethics and the Life Sciences, Hastings Center, Hastings-on-Hudson, New York.

INDEX

GPSR Compliance

The European Union's (EU) General Product Safety Regulation (GPSR) is a set of rules that requires consumer products to be safe and our obligations to ensure this.

If you have any concerns about our products, you can contact us on ProductSafety@springernature.com

In case Publisher is established outside the EU, the EU authorized representative is:

Springer Nature Customer Service Center GmbH
Europaplatz 3
69115 Heidelberg, Germany

Batch number: 10823811

Printed by Printforce, the Netherlands